Strawberry Eats & Treats

THE GUIDE TO ENJOYING STRAWBERRIES

from the
North American
Strawberry Growers
Association

Published by
NORTH AMERICAN STRAWBERRY GROWERS ASSOCIATION
324 Lake Street
Grimsby, Ontario,
Canada L3M 1Z4

Printed by
Palmer Publications, Inc.
PO Box 296
Amherst, WI 54406

Designed and marketed by
Amherst Press
A division of Palmer Publications, Inc.
PO Box 296
Amherst, WI 54406

Contents

 If a boy and girl break open a double **strawberry** *and each eats one half, they will become sweethearts.*

Strawberries

Acknowledgments

The Marketing Committee of the North American Strawberry Growers Association wishes to thank all those who contributed recipes, provided use of graphics and made further contributions toward the production of this book. Numerous sources were consulted and are given credit in the preparation of this book.

We would especially like to give credit to Nourse Farms, Inc., in Deerfield, Massachusetts, and Statz's Berry Land in Baraboo, Wisconsin, for providing photos for the back cover; and Susan Winget and Main Street Press, Delafield, Wisconsin, for the cover art.

White Man celebrates something that happened over 200 years ago. To him, nothing's happened since then. It's all over. All he can do is remember. Indians celebrate what's happening now. When the sacred **strawberries** *come up in the early spring, that's what we celebrate. They're not just* **strawberries** *to us. They're the Creator's gift to his children. They're good to eat, good to drink. But more than that, they have the Creator's power in them. They make us healthy and strong. And we know, when we pass on, that our path to the Sky World will be lined with* **strawberries***. So* **strawberries** *are more than plants to us. That's why every year we have our* **Strawberry** *Thanksgiving— it's something that's happening to us now, not something that happened long ago to somebody else.*

Del Yoder, "Strawberry Thanksgiving,"
Wisdom Keepers: Meeting With Native American Spiritual Elders

Introduction

by Bea Statz, Editor

This book is all about strawberries. It contains information on buying prepicked strawberries, picking your own strawberries, handling strawberries after purchase and recommended methods for freezing strawberries. There are delicious recipes to prepare, savor and enjoy; nutritional information for the health-conscious; low-sugar and low-fat recipes for diabetics; and helpful equivalent measurements to aid in making jam or desserts. Interesting strawberry facts appear throughout the book delighting the reader with poems, quotations and just plain fun things to know about strawberries. If strawberries are not already your favorite fruit, they soon will be.

We've included every possible way to eat and fix strawberries. Each kitchen-tested recipe uses fresh or frozen strawberries or jam as the key ingredient. The chapters include recipes for strawberry muffins, breads, and spreads; dressings, salads, dips; sugarless, low-fat and low-calorie choices; main dishes and entrees; beverages and punches; pies; desserts; and methods for drying strawberries.

Who would be better qualified than strawberry growers to compile a book all about strawberries. Members of the North American Strawberry Growers Association shared their favorite recipes, working knowledge and love of strawberries for this guide.

NASGA members grow and sell their favorite fruit throughout the United States, Canada and 15 foreign countries. Profits from the sale of this book will be used toward research in the production and marketing of strawberries.

 The **strawberry** *crop is the fourth most valuable fruit crop in the United States following grapes, apples and oranges.*

The Strawberry Industry

Strawberry Harvest

by Sally Buckner

Reprinted with permission from *Strawberry Harvest*,
St. Andrews Press, 1986.

May: tangy air whispers a summons;
In dappled light I move to the berry bed.
Leaves green as tomorrow, clusters of triplets
mesh to a lacy coverlet through which sunlight
pours in silver shafts to waiting fruit:
crimson valentines, heavy with juice,
plump as partridges, nesting in yellow straw,
promising jam and shortcake, ready to spurt
at the first touch of tooth. My eager hands
brush cages of leaves and stems, eyes skimming green,
focusing only on scarlet; fingers reach,
touch, pluck, cherish each fat red heart.

At the end of the bed I rise, stretch, pull,
loosen knotted muscles, turn to go,
then stop—caught by flash of ruby, flickering
from a spot I'd swear on a steeple-high pile of Gospels
I examined meticulously five minutes ago,
peering straight down; now the angle of vision
is nearer forty degrees. Intrigued, I review
the entire patch from new perspectives, discover
tasty reward: a dozen more crown jewels
to top the bucket of treasure.

Picking berries requires an agile eye and multiple angles.

NASGA Mission Statement

by Dr. J. W. "Bill" Courter, past Executive Director

The North American Strawberry Growers Association's (NASGA) aim is to promote production and marketing throughout North America by encouraging sustainable culture, supporting basic and applied research, providing educational and marketing programs and acting as an advocate and spokesman for strawberry growers.

NASGA was organized in 1977 and incorporated as a nonprofit corporation by progressive strawberry growers and leading small fruit researchers. Their purpose was to support USDA and state/provincial research programs, develop educational seminars and publications, promote development of equipment, varieties and cultural methods to improve efficiency for the strawberry industry—including grower-applied research and promote beneficial legislation.

Today NASGA represents members in 40 states, 10 provinces in Canada and 15 countries. NASGA continues to be a grower-based association strongly rooted in the original philosophy that ongoing research will provide knowledge to strengthen and improve strawberry production and marketing. To accomplish this goal, NASGA commits 25 percent of dues to research, publishes the journal *Advances for Strawberry Research*, formed a foundation to generate funds for research, publishes a newsletter, develops marketing material for members, sponsors an educational winter conference and supports issues critical to the well-being of strawberry growers through an active legislative committee.

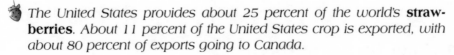

The United States provides about 25 percent of the world's **strawberries**. *About 11 percent of the United States crop is exported, with about 80 percent of exports going to Canada.*

The Strawberry Industry

Buying, Picking, Handling and Freezing Strawberries

Buying Prepicked Strawberries

Prepicked strawberries are available in supermarkets year-round. For this reason, here are some guidelines to consider when purchasing prepicked strawberries. They are sold either by weight or volume. Many times consumers are even allowed to choose each berry individually. By following these tips, buyers can be assured of obtaining a good product.

Sniff before buying. Strawberries, like many fruits, have a wonderful fragrance when fresh and fully ripe; so choose strawberries that are as fragrant and fully ripe as possible. Note that different varieties produce varied shades of red and sweetness. Choose strawberries with a uniform color throughout. Picked berries with green or white tips have little flavor and will not ripen. A good strawberry is also recognized by its sheen and brightness. The fresher the berry, the more they shine. Strawberries become dull and dark when overripe; the flesh should be firm and plump. Select berries that are well-shaped and filled out. The leaves and stem or hull should be fresh, green and well-attached—good signs of a tasty fruit.

When sold by volume (pint, quart or liter), size of the berries is important. Larger berries may not be the best buy since it takes fewer of them to fill a container. Larger berries may also be hollow inside; avoid small berries with seedy bottoms. Try to see the bottom layer to be sure quality and size is uniform throughout. Check the condition of the container before buying. Smell the container for freshness, as it will alert the nose of mildew or mold, which means sour berries. Stains on the container indicate overripe fruit which is losing its juice.

If possible, it is best to buy prepicked strawberries by weight. Not only is it preferred or a better buy, but many times consumers can choose individual berries of the size, color, ripeness and shape they prefer. But keep in mind that berries should not be squeezed and that too much handling causes bruising.

The freshest and best-quality prepicked strawberries to buy are probably those purchased during the fruiting season in the local area. Check first with local strawberry farms, produce stands and commercial farmers' markets for the freshest possible strawberries. Small, high-quality grocery stores may carry local strawberries sooner than large supermarkets.

To properly store fresh-picked strawberries, spread on a cookie sheet or tray to air. Discard spoiled berries and place cookie sheet in refrigerator. Use strawberries as soon as possible. Do not wash or hull until ready to use.

Picking Your Own Strawberries

Strawberries taste best when eaten freshly picked off the vine. Try and pick early in the morning when berries are still cool and the sun hasn't had time to soften them; or second-best, pick in the early evening after the sun has started to set. Check with strawberry growers in your area for the best picking times, conditions, prices, location and availability before driving to local farms. Pick-your-own strawberries can be sold by either weight or volume. Container sizes will vary from farm to farm.

Upon arriving at the farm, customers will be given a container for picking, shown where to park, and assigned a row to pick in. Use both hands to pick berries. In one hand, hold the large branch which supports several berries. With the other hand, pinch stem between thumb and forefinger so that the berry is removed with stem or hull intact. Do not squeeze or pull strawberry, as bruising will damage the fruit.

Strawberries will not appreciably ripen after picking, so be sure each berry is fully ripe with no green or white tips. Keep stems or hulls on fruit. Without them, strawberries will lose their vitamin content, become waterlogged and are vulnerable to mold-causing bacteria.

Pick large and small ripe fruit, avoiding overripe ones. Look for firm, shiny, bright and well-shaped strawberries. Check under all leaves as some of the best fruit is hidden. Keep berries out of the sun, being careful not to fill containers too full or bottom berries will be crushed.

Not only will the fruit be fresher and more nutritious when picking your own, but the cost for superior strawberries will be much less having eliminated picking, handling and shipping costs. It is best to use strawberries as soon as possible. If berries are not used immediately, refrigerate and loosely cover with paper towel or plastic wrap. Do not wash until ready to use.

Handling and Preparing Strawberries

Strawberries are very perishable and should be used or preserved as soon as possible after being purchased or picked. Remove strawberries from containers and sort through and remove any overripe or moldy berries to avoid further contamination. When refrigerating for later use, cover loosely with paper towel or plastic wrap allowing air to circulate to prevent mold. A temperature of 32 to 40 degrees F (0 C to 4 C) is recommended for storage. Strawberries that have been carefully handpicked can be held for two to three days depending on firmness and degree of ripeness.

When ready to use, wash strawberries by carefully placing into a large pan or sink of cool water; remove immediately and drain in a large colander or spread on paper towel. Berries can also be rinsed carefully with cool water. Do not allow strawberries to soak or they will become waterlogged. Hull or remove stem with a small knife or strawberry huller.

Strawberries can be left whole, halved, sliced, mashed, chopped or pureed or used in the method called for in a recipe. Set aside a few whole berries with stem for garnish. Washed and hulled unused berries can be preserved in the refrigerator by adding sugar, orange or lemon juice or adding liqueur, such as Grand Marnier, Cointreau, etc.

Freezing Strawberries

Freezing is recommended for long-term storage of strawberries. The flavor and texture of frozen strawberries are affected by: sweetening agents, amount of sweetener, the method of processing and storage time. Strawberries can be frozen and used with or without sugar. Frozen berries, without sugar, should be used within 3 to 6 months. The flavor, appearance, quality and nutritional value are enhanced when sugar or other sweeteners such as honey is added. Sweetened strawberries can be safely frozen up to one year. The addition of citrus acid and ascorbic acid or lemon or lime juice will also aid appearance and preserve the firmness of frozen strawberries from 3 to 6 months.

Strawberries can be frozen whole, halved, sliced, chopped, mashed or pureed depending upon the desired use later. This is an excellent time to sort berries before freezing: freeze firm and well-shaped berries whole; mash or puree riper, softer fruit and use for making jam, etc. Always label container with exact contents (whole, halved, sliced, chopped, or pureed strawberries), date and amount of sugar or other sweetening added, if any. Thoroughly wash plastic freezer containers and freeze according to amounts called for in recipes. Always work quickly with small amounts of fruit and freeze immediately in coldest part of freezer.

To freeze **whole unsweetened strawberries**—place strawberries in single layer on a cookie sheet; freeze until hard. Quickly transfer to heavy-duty freezer bags or prepared containers and store in freezer. Label; use within 3 to 6 months.

To freeze **whole or sliced unsweetened strawberries** with citrus juice—allow about 1 tablespoon (15 ml) lemon or lime juice to 1 quart (1 L) water. Pack berries in prepared containers and cover with juice. Allow 1/2-inch headspace on top. (When using special ascorbic acid such as FRUIT FRESH from the supermarket, process according to package directions.) Store in freezer. Label; use within 3 to 6 months.

To freeze **whole sweetened strawberries**—allow about 3/4 cup (175 ml) sugar for each 4-5 cups (1.25 L) berries. Gently mix sugar with berries and spread in a single layer on cookie sheets; freeze until hard. Quickly transfer to heavy-duty freezer bags or put directly into prepared containers and store in freezer. Label; use within 1 year.

To freeze **sliced sweetened dry-pack strawberries**—allow about 1 cup (250 ml) sugar for 6 cups (1.5 L) berries. Mix sliced or halved strawberries with sugar and let stand about 10 minutes until sugar dissolves. Place in prepared containers, leaving 1/2-inch headspace. Store in freezer. Label; use within 1 year. Yield: 3 pints = 6 cups berries.

To freeze **sliced or whole syrup-packed strawberries**—This method is considered to be the best freezing medium for quality results. The use of hard water is recommended over soft water. Dissolve about 1 cup (250 ml) sugar in 1-2 cups

(500 ml) water; bring to a boil. Chill to a temperature of 36 degrees F (2 C). Pack strawberries in prepared containers. Pour syrup over berries, leaving 1/2-inch headspace. Store in freezer. Label; use within 1 year.

To freeze **mashed, chopped or pureed unsweetened strawberries**—freeze in small freezer containers as they will take longer to thaw. An alternate method is to pour mixture into ice cube trays, freeze and quickly transfer into heavy-duty freezer bags. This process is excellent for use in beverages, gelatin salads, jams, or bread recipes where sugar is not desired. Store in freezer. Label; use within 6-9 months.

To freeze **mashed, chopped or pureed sweetened strawberries**—allow 1 cup (125 ml) sugar to 4-6 cups (1 L-1.5 L) berries. Stir in sugar until dissolved. Pack into prepared containers of varying sizes, leaving 1/2-inch headspace. Store in freezer. Label; use within 1 year. Serve over ice cream and yogurt, shortcakes or cereal.

To freeze strawberries using a **honey sweetened pack**—replace one half of the sugar with honey. To make syrup, mix 1 cup (250 ml) honey with 3 cups (750 ml) hard (not softened) hot water; bring to a boil. Chill to a temperature of 36 degrees F (2 C). Pack strawberries in prepared containers. Pour syrup over berries leaving 1/2-inch headspace. Store in freezer. Label; use within 1 year.

Extra Special Frozen Strawberries

Strawberries frozen this way provide firm and juicy fruit—
just right for those special-occasion recipes.

Yield: 5 cups (1.25 L)

    3   **quarts (3 L) fresh strawberries, washed and hulled**
 3 1/2 **cups (875 ml) sugar**
   3/4 **cup (175 ml) water**
    1   **box SURE•JELL Fruit Pectin or BALL 100% Natural® Fruit Jell™**

Crush 1 quart (1 L) strawberries to equal 2 cups (500 ml). Add sugar, stir to dissolve. Fill prepared containers with remaining whole, halved or sliced strawberries.

Pour water into small saucepan; slowly add pectin. Bring to a boil; boil 1 minute. Remove from heat and add to crushed mixture. Stir well. Pour mixture over strawberries, leaving 1/2-inch headspace. Store in freezer. Label; use within 1 year.

Strawberry Equivalents and Nutritional Values

Frozen Strawberries:

Commercially unsweetened strawberries are usually sold in 10-ounce (283 g) or 16-ounce (448 g) bags. Measure while still frozen.

Commercially frozen sweetened strawberries are usually sold in 10-ounce (1 1/4-cup) (283 ml) packages.

Fresh Strawberries:

1 quart (1 L) = 1 1/4 to 1 1/2 pounds (675-750 g)

1 quart = 4 cups whole strawberries or 3 cups (750 ml) sliced strawberries or 2 cups (500 ml) pureed strawberries

Yield: 4-5 servings

1 pint (.5 L or 500 ml) = approximately 12 ounces (335-370 g)

1 cup (250 ml) sliced = 1 package (10 ounces) (283 g) of frozen sweetened strawberries

1-1 1/2 quarts (1-1.5 L) = 6 cups (1.5 L) or enough to make a 9-inch (22.5 cm) pie

1 quart (1 L) = 1 batch freezer jam

2 quarts (2 L) = 1 batch cooked jam

8 medium strawberries (5.5) ounces (147 g):

50 calories

1 gram protein

13 grams carbohydrate

0 gram fat

3 grams dietary fiber

0 milligram sodium

Developed by the Food Marketing Institute.
Reviewed by the U.S. Food and Drug Administration.

Strawberry Muffins, Breads, Spreads and More

Humblest born, yet earliest, most abundant and welcome in their season. Everybody ought to eat them; they are available and abundant.

Anonymous (1889)

Straw-Oat Muffins

Use fresh or frozen strawberries

submitted by Bela Casson

*An original recipe, Straw-Oat Muffins
are very nutritious and make great snacks.*

Yield: 12 muffins

1 1/2	cups (375 ml) flour
1	tablespoon (15 ml) baking powder
1/2	teaspoon (2 ml) cinnamon
1/2	teaspoon (2 ml) salt
1/4	cup (50 ml) butter or margarine
1/2	cup (125 ml) sugar
1	egg, beaten
1	cup (250 ml) milk
2/3	cup (175 ml) quick-cooking oats
1/2	teaspoon (2 ml) almond flavoring
1	cup (250 ml) washed, hulled and chopped fresh strawberries or chopped frozen unsweetened strawberries, thawed and drained

Preheat oven to 400 degrees F (200 C); line 12-cup large muffin tin with paper liners or spray with nonstick cooking spray and dust lightly with flour.

Sift together flour, baking powder, cinnamon and salt; set aside.

In a large mixing bowl, cream together butter, sugar and egg. Stir in dry ingredients alternately with milk. Stir in oats. Add almond flavoring. Fold in strawberries.

Spoon into muffin cups, filling three-quarters full. Bake 30 minutes until golden brown or until toothpick inserted in center comes out clean. Brush with melted butter, if desired.

 Strawberry *blossoms are pure white, and like violets, in the shape of little stars. The center is yellow, and this becomes the berry when the petal falls off.*

The Strawberry Connection

Strawberry Banana Muffins

Use fresh or frozen strawberries

submitted by Elsie Maxwell

Elsie Maxwell's original muffin creation is great for breakfast or brunch.

Yield: 12 muffins

2	**large bananas, mashed**
1/2	**cup (125 ml) washed, hulled and chopped fresh strawberries or chopped frozen unsweetened strawberries with juice, thawed**
1	**egg, beaten**
1/3	**cup (75 ml) vegetable oil**
1 1/2	**cups (375 ml) flour**
1/2	**cup (125 ml) sugar**
1	**teaspoon (5 ml) baking soda**
1	**teaspoon (5 ml) baking powder**
1/2	**teaspoon (2 ml) salt**

Preheat oven to 350 degrees F (180 C); line 12-cup muffin tin with paper liners or spray with nonstick cooking spray and dust lightly with flour.

In a medium bowl, mix together bananas, strawberries, egg and oil.

In a separate bowl, sift together flour, sugar, baking soda, baking powder and salt. Add dry ingredients to mashed mixture and stir until blended. Spoon into muffin cups, filling three-quarters full. Bake 20-25 minutes or until toothpick inserted in center comes out clean.

Variation: Use 1 1/2 cups (375 ml) strawberries, omitting bananas or 1/2 cup (125 ml) honey, omitting sugar.

Kenneth and Elsie Maxwell operate a 12-acre vegetable and strawberry farm (plus retail market), on the shores of the Atlantic Ocean in Cape Elizabeth, Maine.

Strawberry Cornmeal Muffins

Use fresh strawberries

submitted by Marilyn Gjevre

This is an easy treat to make during the strawberry season. Strawberry Cornmeal Muffins are great with strawberry jam or other spreads.

Yield: 12 muffins

2	eggs, beaten
1	cup (250 ml) milk
1	teaspoon (5 ml) vanilla extract
1 1/2	cups (375 ml) flour
3/4	cup (200 ml) sugar
3/4	cup (200 ml) cornmeal
1	tablespoon (15 ml) baking powder
1/2	teaspoon (2 ml) salt
1/4	cup (50 ml) butter or margarine, melted
3/4	cup (200 ml) washed, hulled and sliced fresh strawberries

Preheat oven to 400 degrees F (200 C); line 12-cup muffin tin with paper liners.

Combine eggs, milk and vanilla in small mixing bowl, beat thoroughly.

In separate large bowl, combine flour, sugar, cornmeal, baking powder and salt. Stir egg mixture into flour mixture; add butter, stirring just until mixed. Do not overmix as batter will be lumpy.

Fold in strawberries.

Fill each baking cup three-quarters full. Bake 15 minutes or until toothpick inserted in center comes out clean.

Paul and Marilyn Gjevre of Sweetwater Farm, four miles east and two miles south of Rosholt, South Dakota, have pick-your-own strawberries.

Strawberry Sour Cream Bread

Use fresh or frozen strawberries

submitted by Debbie Lineberger

This moist quick bread is particularly meant for those who, like the Linebergers, "prefer fruit breads better than cakes."

Yield: 1 loaf

2 1/3	**cups (575 ml) Bisquick or Jiffy Baking Mix**
3/4	**cup (200 ml) sugar**
1/3	**cup (75 ml) sour cream**
1/4	**cup (50 ml) vegetable oil**
1	**teaspoon (5 ml) cinnamon**
3	**eggs**
2	**teaspoons (10 ml) vanilla extract**
1	**cup (250 ml) washed, hulled and chopped fresh strawberries or chopped frozen unsweetened strawberries, partially thawed**
1/2	**cup (125 ml) nuts, chopped**

Preheat oven to 350 degrees F (180 C). Spray a 9x5x3-inch (22.5x12.5x7.5 cm) loaf pan with nonstick cooking spray and dust lightly with flour.

Combine baking mix, sugar, sour cream, oil, cinnamon, eggs and vanilla; beat 50 strokes by hand. Fold in strawberries and nuts.

Pour into prepared pan and bake 45-60 minutes until toothpick inserted in center comes out clean. Cool loaf 5-10 minutes on wire rack before removing from pan.

When cool wrap in plastic wrap or aluminum foil and store in refrigerator.

 We are bound by a small, sometimes magical fruit called the **strawberry**. *This fruit has the power to make tears dry up, make friends with enemies, make sick people feel better, make the elderly feel younger by bringing back pleasant memories of days gone by, make acquaintances of strangers, and above all make little children smile. What other fruit has that power?*

Marvin Brown, "Welcome Speech," 1995 NASGA Conference

Strawberry Coffee Cake

Use fresh strawberries

submitted by Mary Secor

*The credit for this delicious coffee cake goes to Mary Secor's mother-in-law.
It's great for breakfast, brunch or anytime.*

Yield: 1 cake

   1/2 **cup (125 ml) sugar**
     1 **cup (250 ml) flour**
     2 **teaspoons (10 ml) baking powder**
  1/2 **cup (125 ml) milk**
     1 **egg, beaten**
     2 **tablespoons (25 ml) butter or margarine, melted**
1 1/2 **cups (375 ml) washed, hulled and sliced fresh strawberries**

Topping:
  1/2 **cup (125 ml) flour**
  1/2 **cup (125 ml) sugar**
  1/4 **cup (50 ml) butter, softened**
  1/4 **cup (50 ml) chopped nuts**

Preheat oven to 375 degrees F (190 C). Spray an 8-inch (20 cm) square baking pan with nonstick cooking spray and set aside.

In large mixing bowl, combine sugar, flour, baking powder, milk, egg and butter. With electric mixer on medium, beat 2 minutes until well blended. Pour batter into prepared pan; arrange strawberries evenly over batter.

To make topping, in a medium bowl combine flour and sugar. With pastry blender or two knives used scissor-fashion, cut in butter until mixture resembles coarse crumbs. Stir in nuts. Sprinkle topping evenly over strawberries. Bake 35-40 minutes.

Serve warm or cooled.

 Donald and Mary Secor have operated Secor Strawberries Inc., for over thirty years. Located in the mid-Hudson Valley halfway between New York City and Albany, New York, they sell pick-your-own strawberries, peas and pumpkins.

Strawberry Buckle

Use fresh strawberries

submitted by Marilyn Gjevre

Marilyn Gjevre receives great reviews when Strawberry Buckle is served during coffee break. This versatile recipe can also be served for breakfast, brunch, dessert or as a snack.

Yield: 1 cake (9 inches) (22.5 cm)

1/4	cup (50 ml) butter or margarine, softened
3/4	cup (200 ml) sugar
1	egg, beaten
1/2	teaspoon (2 ml) vanilla extract
1/2	cup (125 ml) milk
2	cups (500 ml) cake flour, sifted
2	teaspoons (10 ml) baking powder
1/2	teaspoon (2 ml) salt
2	cups (500 ml) washed, hulled and sliced fresh strawberries

Topping:

1/4	cup (50 ml) butter, softened
1/2	cup (125 ml) brown sugar
1/3	cup (75 ml) cake flour, sifted
1/2	teaspoon (2 ml) cinnamon
	Dash of nutmeg

Preheat oven to 375 degrees F (190 C). Spray a 9-inch (22.5 cm) square pan with nonstick cooking spray and dust lightly with flour. In a large mixing bowl, cream butter and sugar until light and fluffy. Blend in egg and vanilla, mixing well. Add milk, mixing until blended.

In a separate bowl, sift together flour, baking powder and salt. Stir into liquid mixture. Fold in strawberries. Spread into pan.

To make topping, in a small bowl cream together butter and sugar. Blend in flour and cinnamon. Sprinkle mixture over batter and top with dash of nutmeg. Bake 30-35 minutes.

Serve warm with cream, whipped topping or ice cream.

Strawberry Butter

Use fresh strawberries

submitted by Doris Stutzman

Strawberry Butter is delicious spread on pancakes, muffins, bagels or toast.

Yield: 1 3/4 cups (450 ml)

- 3/4 **cup (200 ml) fresh whole strawberries, washed and hulled**
- 3/4 **cup (200 ml) butter, softened**
- 1 **tablespoon (15 ml) honey**

Bring strawberries, butter and honey to room temperature before mixing.

Puree strawberries in blender. Whisk berries gradually into butter and honey until well blended. Serve immediately or refrigerate until ready to use.

Cream Cheese Strawberry Spread

Use frozen strawberries

submitted by Bea Statz

This recipe makes a smooth pink spread for bread, bagels or muffins.

Yield: 1 cup (250 ml)

1 package (10 ounces) (283 g) frozen sweetened strawberries, thawed
1 package (8 ounces) (240 g) cream cheese, softened

Drain and reserve juice from strawberries. Place 1/2 cup (125 ml) strawberry juice and cream cheese in blender. Process until spreading consistency.

Note: The remaining strawberries and juice can be used for making strawberry bread, muffins, gelatin salads, etc.

 The **strawberry** *is the symbol of perfect righteousness. Not only is the* **strawberry** *a religious symbol, but it dates back to a relatively early origin.*

The Strawberry Connection

Strawberry Dressings, Salads and More

For centuries **strawberries** were planted as ground covers in beds of flowering shrubs. In 1368, 12,000 plants were purchased for the Royal Gardens of the Louvre. Wild **strawberry** plants were cultivated in the Garden of Versailles.

The Strawberry Connection

Strawberry Vinegar

Use fresh strawberries

submitted by Susan Butler

This slightly sweet strawberry vinegar is excellent for salads and also as a glaze to complement pork, chicken or beef.

Yield: 2 cups (500 ml)

1 1/2   **cups (375 ml) washed and hulled small ripe strawberries**
1 1/2   **cups (375 ml) white wine vinegar**
 1/2   **cup (125 ml) sugar**

Place strawberries and vinegar in a 1-quart (1 L) jar. Seal jar and let stand in a sunny window 3-4 days (vinegar should be bright red).

Strain vinegar into a small stainless steel or enamel saucepan; discard berries. Stir in sugar and heat over medium heat until mixture comes to a boil. Simmer over low heat approximately 10 minutes. Cool and pour into pint bottle. Cover and store in a cool, dry place or in refrigerator.

Tossed Salad with Strawberry Vinegar Dressing

Use fresh strawberries

submitted by Susan Butler

Strawberry Vinegar is the base for this excellent tossed salad dressing.

Yield: 3/4 cup (200 ml)

Dressing:
 1/2   **cup (125 ml) Strawberry Vinegar**
    (see above recipe)
 1/4   **cup (50 ml) extra light olive oil**
  1   **teaspoon (5 ml) Worcestershire sauce**
  1   **fresh clove garlic, minced**
  2   **whole cloves**

Salad:
Fresh spinach
Tomato wedges
Mushrooms
Parmesan cheese
Croutons

Combine and mix Strawberry Vinegar, oil, Worcestershire, garlic and cloves. Marinate in refrigerator for 24 hours; remove and discard cloves. Refrigerate unused dressing.

Prepare salad ingredients and toss with dressing just before serving.

Variation: Substitute any lettuce or combination of lettuces for spinach.

Strawberry Lettuce Salad

Use fresh strawberries

submitted by Bill and Treva Courter

The friends and neighbors of Bill and Treva Courter in Kentucky have a hard time eating a small portion of their favorite tossed salad.

Yield: 12 servings

Salad:

    Leaf Lettuce, washed and torn
    2 cups (500 ml) washed, hulled and sliced fresh strawberries

Dressing:

    3/4 cup (200 ml) sugar
    1/2 cup (125 ml) red wine vinegar
    1/4 cup (50 ml) extra virgin olive oil
    1 teaspoon (5 ml) paprika
    2 fresh cloves garlic, minced
    1/2 teaspoon (2 ml) pepper

In a microwave-proof bowl, blend together sugar, vinegar, oil, paprika, garlic and pepper. Heat until dissolved and blended. Can be prepared ahead and served warm or cold.

Arrange sliced strawberries over lettuce in large bowl and pour over dressing. Toss slightly.

Bill and Treva Courter served as executive secretaries of NASGA from 1993-1996. Both grew up on strawberry farms. Bill, professor emeritus, University of Illinois, is a charter member of NASGA and has worked as director, program chair and researcher in consumer marketing.

Strawberry Lettuce Salad with Poppy Seed Dressing

Use fresh strawberries

submitted by Cherie Reilly

This combination of green lettuce, sweet onions, strawberries, oranges and poppy seeds is pleasing to the eye and tempting to the palate.

Yield: 8 servings

Iceberg lettuce or a variety of your choice
1   medium Vidalia or Bermuda onion
1   can (11 ounces) (312 g) mandarin oranges, drained
1   pint (500 ml) fresh strawberries, washed and hulled (sliced if large)

Dressing:
3/4   cup (200 ml) sugar
1   teaspoon (5 ml) dry mustard
1   teaspoon (5 ml) salt
1/3   cup (75 ml) cider vinegar
2   teaspoons (10 ml) chopped scallions
1   cup (250 ml) vegetable oil
1 1/2   tablespoons (22 ml) poppy seeds

Tear lettuce into bite-size pieces; wash and drain. Peel onion and cut in very thin slices. In a large glass bowl, add lettuce, onions, oranges and strawberries.

To make dressing, in a blender, mix sugar, mustard, salt and vinegar until smooth. Add scallions; continue to blend. Add vegetable oil slowly while blender is running. Blend until dressing is thick. Stir in poppy seeds.

Pour dressing over salad mixture and toss.

Michael and Cherie Reilly's Summer Seat Farm is nestled in the hills along the Ohio River just 15 minutes from Pittsburgh, Pennsylvania. Their main crops are you-pick strawberries and pumpkins.

Spinach Strawberry Walnut Salad

Use fresh strawberries

submitted by Doris Stutzman

*A warm dressing makes this colorful and nutritious salad
perfect for any occasion.*

Yield: 8 servings

Salad:
- 1 1/2   pounds (750 g) fresh spinach, washed
- 3   cups (750 ml) fresh strawberries, washed and hulled
- 1   large red onion, sliced thin
- 1   cup (250 ml) walnuts

Dressing:
- 2   tablespoons (25 ml) cornstarch
- 2   tablespoons (25 ml) sugar
- 1   cup (250 ml) water
- 3   tablespoons (45 ml) margarine
- 1/2   cup (125 ml) Strawberry Vinegar (see page 18)

To make salad, arrange spinach, strawberries, onion and walnuts in a small bowl. Refrigerate until ready to serve.

To make dressing, mix together cornstarch, sugar and water in medium saucepan. Bring to a boil until mixture thickens. Add margarine and stir until melted. Add vinegar; stir to combine. Keep warm until ready to serve.

Serve warm dressing over salad or on the side.

Variation: Substitute dried strawberries for fresh in the off-season.

 For more than 150 years, the Stutzman family of Indiana, Pennsylvania, has earned a reputation as growers of quality strawberries, fruits and vegetables. The Berry Hill kitchen transforms fresh fruit into jams, jellies and other products that are sold in gift baskets.

Strawberry Pretzel Salad

Use frozen strawberries

submitted by Cheryl Halat

*Served as a salad or dessert, this tasty combination
is a great way to use frozen strawberries.*

Yield: 12 servings

Layer 1: (Crust)
- 2   cups (500 ml) coarsely ground pretzels
- 1½   sticks (¾ cup) (200 ml) butter or margarine, melted
- 2   tablespoons (25 ml) sugar

Layer 2: (Filling)
- 1   package (8 ounces) (226 g) cream cheese
- 1   cup (250 ml) sugar
- 1   container (16 ounces) (454 g) whipped topping

Layer 3: (Topping)
- 2   cups (500 ml) boiling water
- 1   package (6 ounces) (170 g) strawberry gelatin
- 2   packages (10 ounces each) (283 g each) frozen sweetened strawberries or 2 cups (500 ml) frozen unsweetened strawberries, partially thawed

To make topping, mix boiling water and strawberry gelatin; stir until dissolved. Add strawberries and refrigerate for about 30 minutes or until mixture thickens but is not set.

Preheat oven to 375 degrees F (190 C).

To make crust, in a small bowl, mix pretzels, butter and sugar. Pat mixture into a 9x13-inch (22.5x30.5 cm) baking pan. Bake 10 minutes. Remove from oven; cool.

To make filling, in a medium bowl, cream together cream cheese and sugar until thoroughly blended. Fold in whipped topping. Spread over cooled crust.

Spoon strawberry-gelatin mixture over cream cheese layer. Refrigerate until firm.

The Tom and Cheryl Halat family of Huntley, Illinois, operate three farm markets besides growing strawberries and other vegetables.

Strawberry Rhubarb Gelatin Salad

Use fresh or frozen strawberries

submitted by Carolyn Beinlich

*This colorful jello mold blends rhubarb and strawberry flavors.
Both fruits are available fresh during the spring or from the freezer in winter.*

Yield: 8 servings

4	**cups (1 L) diced raw rhubarb**
1 1/2	**cups (375 ml) water**
1/2	**cup (125 ml) sugar**
1	**large package (6 ounces) (170 g) strawberry gelatin**
1	**cup (250 ml) orange juice**
1	**tablespoon (15 ml) orange rind**
2	**cups (500 ml) washed, hulled and sliced fresh strawberries; reserve whole strawberries with hulls for garnish**

Combine rhubarb, water and sugar in saucepan and cook until tender.

In a large bowl, pour hot rhubarb mixture over gelatin and stir until completely dissolved. Add orange juice and rind. Refrigerate until syrupy, about 45 minutes.

Fold fresh strawberries into rhubarb mixture. Pour into 1-quart (1 L) mold and refrigerate until set.

Unmold and garnish with whole fresh strawberries.

Note: If using frozen strawberries for fresh, add them frozen to hot rhubarb with orange juice and rind. Stir often to be sure strawberries are thawed before gelatin is set.

Variation: Substitute frozen rhubarb for fresh.

 The ripe **strawberries** *quench thirst and take away, if they be often used, the rednesse and heat of the face.*

Gerard's Herball (1597)

Mrs. Burn's Layered Strawberry Salad

Use fresh or frozen strawberries

submitted by Helen Thoren Kent

This salad is always a highlight at church suppers. Rev. Burns' wife is credited for it whenever Helen Thoren Kent brings it to the covered-dish suppers.

Yield: 12 servings

1	package (6 ounces) (170 g) strawberry gelatin
1 1/2	cups (375 ml) boiling water
1	package (10 ounces) (283 g) frozen sweetened strawberries and juice, partially thawed
1	can (20 ounces) (567 g) crushed pineapple with juice
1	large banana, mashed
1	cup (250 ml) sour cream

Dissolve gelatin in boiling water. Stir strawberries, pineapple and banana into gelatin. Pour one half of gelatin-fruit mixture into 9x13-inch (22.5x30.5) flat dish and refrigerate until partially set (about 1 hour).

Spread sour cream over gelatin-fruit layer. Pour remaining gelatin-fruit mixture over sour cream layer and refrigerate until firm.

Variation: Substitute 1 cup (250 ml) washed, hulled and sliced fresh strawberries for frozen.

 Five generations of Kents have operated Locust Grove Farm in Milton, New York, along the lower Hudson River, about 80 miles north of New York City. Along with five acres of strawberries, which are sold both you-pick and prepicked, they grow 40 varieties of apples; peaches, pears, plums, grapes, currants, quinces, blackberries, raspberries and blueberries.

Strawberry Dips

A ripe **strawberry** could be compared to the fruit of the rose (the hip, full of seeds) turned inside out.

The Strawberry Connection

Easy Dipping Ideas

Use fresh strawberries

submitted by Carolyn Beinlich

This variety of easy and simple dips can be ready in no time and served as an appetizer, snack or dessert.

Yield: 12 servings

Powdered sugar
Sour cream followed by brown sugar
Plain yogurt followed by brown sugar
Melted white or semisweet chocolate
Flavored yogurt
1 **quart (1 L) fresh large whole strawberries with hulls, washed and air-dried on paper towel**

Place each ingredient in individual small bowls. Arrange strawberries in a larger bowl.

Yogurt Dip

Use fresh strawberries

submitted by Diane Wenham

Impress your guests with this delicious smooth-dipping treat.

Yield: 1 cup (250 ml)

1 **cup (250 ml) vanilla yogurt**
1 **tablespoon (15 ml) lemon juice**
2 **tablespoons (25 ml) liquid honey**
1/2 **teaspoon (2 ml) almond flavoring**
1 **quart (1 L) fresh whole perfect strawberries with hulls and caps intact, washed and air-dried on paper towel**

In a small mixing bowl, combine yogurt, lemon juice, honey and almond flavoring. Blend on medium speed with electric mixer until smooth. Refrigerate.

Arrange strawberries in a bowl over crushed ice or on a platter. Serve dip in separate bowl.

Chocolate Mint Dip

Use fresh strawberries

submitted by Bea Statz

A special light dip sure to please the chocolate lover—a sweet complement with appetizers, hors d'oeuvres or snacks. This recipe was furnished by the California Strawberry Advisory Board.

Yield: 3/4 cup (200 ml)

1/2 **cup (125 ml) nonfat sour cream**
1/4 **cup (50 ml) light chocolate fudge sauce**
3-4 **drops peppermint extract**
1 **quart (1 L) fresh large whole strawberries with hulls, washed and air-dried on paper towel**

Whisk sour cream, fudge sauce and peppermint extract together; blend thoroughly.

Serve strawberries and dip in separate bowls.

Marshmallow Cream Cheese Dip

Use fresh strawberries

submitted by Carolyn Beinlich

Try this smooth, creamy dip with a variety of flavors.

Yield: 2 cups (500 ml)

1 **package (8 ounces) (226 g) cream cheese, softened**
1 **jar (7 ounces) (198 g) marshmallow creme**
1 **quart (1 L) fresh large whole strawberries with hulls, washed and air-dried on paper towel**

Whip cream cheese and marshmallow creme on medium speed with electric mixer until fluffy. Add flavoring of your choice, if desired.

Serve dip and strawberries in separate bowls.

Optional flavoring variations: Add grated lemon peel and 2 teaspoons (10 ml) lemon juice; add grated orange peel and 2 teaspoons (10 ml) orange juice; add 2 tablespoons (25 ml) brown sugar and 1 tablespoon (15 ml) Kahlua coffee liqueur; add 1 teaspoon (5 ml) almond extract.

Tangy Fruit Dip

Use fresh strawberries

submitted by Carolyn Beinlich

Try this tangy dip for any fruit plate that includes strawberries.

Yield: 2 cups (500 ml)

1   **teaspoon (5 ml) honey**
1 1/2   **cups (375 ml) frozen whipped topping, thawed or whipped cream**
1/2   **cup (125 ml) grenadine syrup**
1/2   **cup (125 ml) mayonnaise**
1   **quart (1 L) fresh large whole strawberries with hulls,
     washed and air-dried on paper towel**

In small bowl, combine honey, whipped topping, grenadine, and mayonnaise. Blend on medium speed with electric mixer until well mixed. Refrigerate until ready to serve. Serve whole strawberries and dip in separate bowls.

Spice Dip

Use fresh strawberries

submitted by Carolyn Beinlich

The spicy flavor of this dip makes it a real winner!

Yield: 1 2/3 cups (375 ml)

1/4   **cup (50 ml) firmly packed brown sugar**
1/8   **teaspoon (.5 ml) cinnamon**
     Dash of nutmeg
2   **cups (500 ml) frozen whipped topping, thawed**
1   **quart (1 L) fresh large whole strawberries with hulls,
     washed and air-dried on paper towel**

In a small bowl, mix brown sugar with cinnamon and nutmeg. Fold into whipped topping. Garnish with additional nutmeg or cinnamon, if desired. Refrigerate until ready to serve.

Serve strawberries and dip in separate bowls.

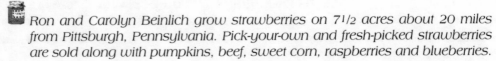 *Ron and Carolyn Beinlich grow strawberries on 7 1/2 acres about 20 miles from Pittsburgh, Pennsylvania. Pick-your-own and fresh-picked strawberries are sold along with pumpkins, beef, sweet corn, raspberries and blueberries.*

Strawberry Fondue

Use frozen strawberries

submitted by Carolyn Beinlich

*Try a cooked strawberry fondue with a variety of party-dipping ingredients;
it's appealing and delicious.*

Yield: 3 cups (750 ml)

2	**packages (10 ounces each) (283 g each) frozen sweetened strawberries with juice, partially thawed**
1/4	**cup (50 ml) cornstarch**
1/2	**cup (125 ml) water**
1	**package (3 ounces) (85 g) cream cheese, softened**
1/4	**cup (50 ml) brandy (optional)**

**Fresh whole strawberries with hulls, washed
and air-dried on paper towel**

Crush strawberries slightly in saucepan. In a small bowl, blend together cornstarch and water. Add to berries and cook over medium heat, stirring constantly until berries are broken and mixture is thick and bubbly. Remove from heat. Blend in cream cheese, stirring until melted. Gradually add brandy, stirring constantly. Pour into fondue pot.

Serve with whole strawberries, fresh or canned fruits (peaches, pears, pineapple), marshmallows or cake cubes.

Variation: Use unsweetened strawberries, adding 2 tablespoons (25 ml) sugar. Substitute a drop of vanilla, almond or lemon extract instead of brandy.

 *The witty and delightful Anne Boleyn had a **strawberry**-shaped birthmark on her slender neck—to her eventual dismay. Her enemies said this mark was sure evidence that the queen was a witch.*

Strawberries

Chocolate Dipped Strawberries in a Basket

Use fresh strawberries

submitted by Mary Secor

Serve this elegant basket of strawberries from Womens World *magazine for special meetings and when friends come to visit.*

Yield: 12 servings

6 **ounces (177 g) white candy melts or white chocolate**
4 **ounces (125 g) semisweet chocolate chips or bulk semisweet chocolate**
1 **pint (2 cups) (500 L) fresh whole perfect strawberries, washed, hulled and air-dried on paper towel**

Melt white chocolate over low heat in saucepan, double boiler or microwave. Spoon into heavy plastic bag. Cool slightly for 5 minutes.

Meanwhile, place aluminum foil over inside of 2¹/₂-cup (625 ml) bowl. Spray inside with nonstick cooking spray. Cut a small hole in one corner of a plastic bag; drizzle white chocolate onto bottom and sides of bowl in a lattice pattern to form a fairly thick basket. Place bowl in freezer for 30 minutes or until ready to serve.

Line cookie sheet with waxed paper; set aside.

Melt semisweet chocolate in saucepan, double boiler or microwave. Dip strawberries in chocolate one at a time, leaving the top one-quarter of berry uncovered. Shake off excess. Place dipped berries on prepared pan. Chill until firm, about 15 minutes.

Just before serving, remove basket and foil from bowl. Unmold chocolate by carefully peeling away foil. Arrange basket on serving plate; place dipped strawberries inside.

 Consumers select **strawberries** *based mainly on four characteristics: color, flavor, shape and size. But for repeat purchases, the majority of retailers agree that flavor is the number one.*

Produce Business

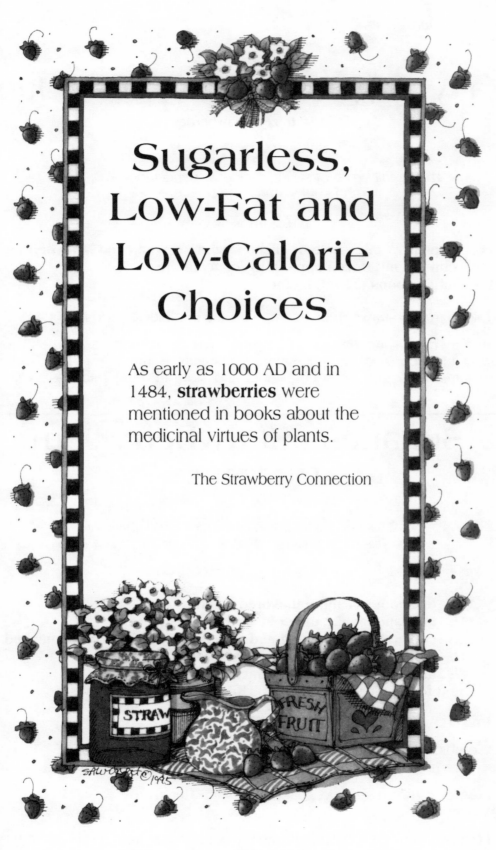

Sugarless, Low-Fat and Low-Calorie Choices

As early as 1000 AD and in 1484, **strawberries** were mentioned in books about the medicinal virtues of plants.

The Strawberry Connection

Easy Sugarless Strawberry Jam

Use frozen strawberries

submitted by Bea Statz

*This small batch of sugarless jam is perfect for the diabetic
who cooks small amounts of food.*

Yield: 2 cups (500 ml)

- 1 1/2 **cups (375 ml) whole or sliced frozen unsweetened strawberries**
- 1/4 **cup (50 ml) frozen apple juice concentrate**
- 1 1/2 **tablespoons (22 ml) tapioca**

Blend frozen strawberries and apple juice in a blender; do not puree.

In a medium saucepan, add mixture to tapioca; let stand 5 minutes. Slowly bring to a boil on medium heat, stirring constantly. Remove from heat and cool 20 minutes. Pour into clean containers and store in refrigerator. Use within 3 weeks.

Sugarless Strawberry Jam

Use fresh strawberries

submitted by Elaine Maust

*This is a larger batch of jam for the diabetic.
It calls for fresh strawberries and can be frozen.*

Yield: 3 cups (750 ml)

- 1 1/2 **teaspoons (7 ml) unflavored gelatin**
- 1 1/2 **tablespoons (22 ml) cold water**
- 3 **cups (750 ml) washed and hulled fresh strawberries, mashed**
- 1 1/2 **tablespoons (22 ml) liquid sweetener**
- 1/4 **teaspoon (1 ml) ascorbic acid powder**
- **Red food coloring, if desired**

In a cup, soften gelatin with water and set aside.

Combine mashed strawberries and sweetener in a medium saucepan. Place over high heat, stirring constantly until mixture comes to a boil.

Remove from heat; add softened gelatin. Return to heat and continue to cook 1 minute.

Remove from heat. Blend in ascorbic acid powder and food coloring. Pour into freezer containers. Cover. Store in refrigerator or freezer.

Low-Fat Strawberry Bread

Use fresh strawberries

submitted by Linda Stanley-Ramos

Yield: 2 loaves

1/2	cup (125 ml)	chopped walnuts or pecans (optional)
1	cup (250 ml)	sugar
1/4	cup (50 ml)	brown sugar
1	cup (250 ml)	egg whites or fat-free egg substitute
1/2	cup (125 ml)	unsweetened applesauce
1/2	cup (125 ml)	buttermilk
1	teaspoon (5 ml)	vanilla extract
1/4	teaspoon (1.5 ml)	almond extract
3	cups (750 ml)	unbleached flour, divided
1	teaspoon (5 ml)	cinnamon
1	teaspoon (5 ml)	baking soda
1/4	teaspoon (1 ml)	salt
3	cups (750 ml)	washed, hulled and sliced fresh strawberries

Preheat oven to 350 degrees F (180 C). Lightly coat two 9x5x3-inch (22.5x12.5x7.5 cm) loaf pans with nonstick cooking spray.

Lightly toast nuts on cookie sheet 10-12 minutes; set aside to cool.

In a bowl, combine sugar, brown sugar, egg whites, applesauce and buttermilk. Using an electric mixer, beat on medium speed until the sugar dissolves and mixture is blended. Add vanilla and almond extracts and beat 1 minute longer.

In a separate bowl, combine flour, cinnamon, baking soda and salt. Stir with fork or whisk to blend. Remove 1 tablespoon (15 ml) of flour mixture and sprinkle over sliced berries; gently stir, coating berries with flour to prevent them from sinking to bottom of loaf.

At low speed, beat half of the flour mixture into sugar-applesauce mixture. Sift in remaining flour alternately with sliced berries and nuts. Batter may appear lumpy. Pour batter into prepared pans, filling each three-quarters full. Thump bottoms of pans firmly on counter to force out air bubbles.

Bake 50-60 minutes or until a toothpick inserted in center of loaf comes out clean. If the top is browning too quickly, cover loosely with foil. Allow loaves to cool on wire rack 15-20 minutes before removing from pans. When completely cool, wrap in plastic wrap or aluminum foil and store in refrigerator.

Note: Do not substitute frozen berries or bread will be soggy.

 Linda Stanley-Ramos of Produce Promotions in Illinois, has produced sales advertising and marketing materials for NASGA members for 15 years.

Strawberry Summer Soup

Use fresh or frozen strawberries

submitted by Bela Casson

A summer favorite, this cold soup is elegant as an appetizer or dessert.

Yield: 1 quart (1 L)

1   **quart (1 L) fresh strawberries, washed and hulled**
    or 4 cups (1 L) frozen unsweetened whole strawberries
1/3   **cup (75 ml) sugar**
2   **tablespoons (25 ml) honey**
1   **cup (250 ml) strawberry yogurt**
1   **teaspoon (5 ml) lemon juice**
1/3   **cup (75 ml) water**
1/3   **cup (75 ml) port wine**
Dash of nutmeg

Blend strawberries, sugar, honey and yogurt in a food processor or blender until smooth. Add lemon juice, water, wine and nutmeg. Blend thoroughly. Pour into a bowl and refrigerate several hours to blend flavors.

Serve very cold, garnished with mint leaf or twist of lemon.

Variation: Substitute artificial sweetener for sugar and honey; non-alcoholic wine for port wine.

*In 1995, 110 **strawberry** growers in Nova Scotia harvested 3.8 million quarts from approximately 750 acres. Thirty-five percent was you-picked, 60 percent sold to retail, and 5 percent went to processing.*

Strawberry Growers of Nova Scotia
& Nova Scotia Federation of Agriculture

Low-Fat Cheesecake Dip

Use fresh strawberries

submitted by Carolyn Beinlich

This dip may be low in fat, but it is high in flavor. Guests will want this recipe.

Yield: 3 cups (750 ml)

1   **package (3 ounces) (85 g) light neufchatel cheese, softened**
2   **tablespoons (25 ml) sugar or equivalent artificial sweetener**
3   **tablespoons (45 ml) low-fat milk**
2   **cups (500 ml) light whipped topping**
1   **teaspoon (5 ml) vanilla extract**
**Fresh large whole strawberries hulled, washed and air-dried
    on paper towel**

Blend neufchatel cheese, sugar and milk until well blended and smooth. Fold in whipped topping and vanilla. Put in a serving dish surrounded by dipping strawberries.

Variation: Substitute orange or lemon juice for vanilla.

 *Some historians feel that the tenderness found in 14th century paintings and prayer books, including sketches of **strawberries**, was a reflection of the gentle Saint Francis of Assisi who loved all of God's creation.*

The Strawberry Connection

Sugarless Strawberry Fruit Spread

Use fresh strawberries

submitted by Bea Statz

Serving ideas for this SURE•JELL recipe for sugarless spread include: mix with plain yogurt; spread on angel food or fat-free pound cake as a filling for a low fat dessert; mix with a little orange juice to make a glaze for ham, chicken or pork; sandwich between vanilla wafers, sugar cookies or ginger snaps; use a thin layer on the bottom of tarts.

Yield: 5 cups (1.25 L)

2   **quarts (2 L) fresh fully ripe strawberries, washed and hulled = 4 cups mashed (1 L)**
2   **cans (12 ounces each) (355 ml each) frozen concentrated white grape juice, thawed**
1   **box SURE•JELL for Lower Sugar Recipes Fruit Pectin**
1/2 **teaspoon (2 ml) butter or margarine**

Thoroughly mash strawberries, mashing one layer at a time. Measure 4 cups (1 L) into a 6- to 8-quart-deep saucepan. (Do not use a smaller saucepan or fruit spread will not set.) Stir grape juice into strawberries.

Wash and sterilize jars. Keep jars and lids hot in boiling water until ready to fill; drain.

Stir pectin gradually into fruit mixture. Add butter. Place over high heat; bring to a full rolling boil, stirring constantly. Continue boiling over high heat 20 minutes, stirring constantly. Remove from heat; skim off foam with metal spoon.

Quickly ladle into prepared jars, filling to within 1/8 inch of tops. Wipe jar rims and threads. Cover with two-piece lids. Screw band tightly. Invert jars 5 minutes or process with boiling water-bath. Check to see that all jars have sealed.

Allow 1 week for spread to set before using.

 Roman and Bea Statz of Baraboo, Wisconsin, have grown pick-your-own and fresh-picked strawberries on 10 to 12 acres since 1977. Bea is editor of **Strawberry Eats & Treats** *and has previously published a book of strawberry recipes. She serves on the NASGA Marketing Committee and chairs the Wisconsin Berry Recipe Contest.*

Strawberry-Orange Allspice Topping

Use fresh strawberries

submitted by Linda Stanley-Ramos

Yield: 2 cups (500 ml)

1/4   cup (50 ml) orange juice
2   tablespoons (25 ml) sugar or equivalent artificial sweetener
1/2   teaspoon (2 ml) allspice
2   pints (4 cups) (1 L) fresh strawberries, washed,
    hulled and halved
1   teaspoon (5 ml) grated orange peel

Combine orange juice, sugar and allspice in small bowl; stir until sugar dissolves. Put the strawberries in individual dessert dishes or a large glass bowl. Pour mixture over strawberries and sprinkle with grated orange peel.

Strawberry-Pineapple Yogurt Delight

Use frozen strawberries

submitted by Elaine Maust

Yield: 6 servings

1   envelope unflavored gelatin
1   cup (250 ml) unsweetened crushed canned pineapple, drained;
    reserve juice
3   cups (750 ml) frozen unsweetened strawberries,
    thawed and undrained
1   cup (250 ml) low-fat strawberry yogurt
1   teaspoon (5 ml) grated lemon peel
Low-fat whipped topping (optional)
Lemon wedges

Sprinkle gelatin over reserved pineapple juice in small saucepan. Let stand 1 minute. Heat on low 2-3 minutes or until gelatin dissolves, stirring constantly. Remove from heat.

Add strawberries, yogurt and lemon peel; mix well. Pour into individual dessert dishes. Refrigerate. Garnish with whipped topping and lemon wedge before serving.

Frozen Strawberry-Orange Bars

Use fresh strawberries

submitted by Jo Gibbs

High in Vitamin C and containing zero fat, this recipe is a good snack for kids and one they'll love making and eating. This is an official 5-a-day recipe.

Yield: 4 bars (4 ounces each) (120 g each)

1 **cup (250 ml) fresh whole strawberries, washed and hulled**
1 **cup (250 ml) orange sections**
1 **tablespoon (15 ml) sugar or equivalent artifical sweetener (optional)**
1 **teaspoon (5 ml) lemon juice**
4 **wooden popsicle sticks**

Puree strawberries and orange sections in blender, adding a tablespoon or two (15-25 ml) of water, if necessary. Add sugar and lemon juice; blend. Pour into bar molds or small cups and insert popsicle sticks. Freeze until solid.

Variation: Substitute banana for orange sections.

Strawberry Smoothie & Frozen Pops

Use fresh strawberries

submitted by Elaine Maust

This original beverage recipe has the health conscious in mind and can be frozen into popsicles.

Yield: 2 cups (500 ml) Smoothies or 2-4 popsicles

1/2 **cup (125 ml) washed, hulled and sliced fresh strawberries**
1/3 **banana**
1/4 **cup (50 ml) vanilla or plain yogurt**
1/4 **cup (50 ml) orange juice**
1 **ice cube**

Combine strawberries, banana, yogurt, orange juice and ice cube in blender or food processor. Blend until smooth.

Pour into glasses and garnish with a whole strawberry with stem.

To make pops, fill 2-4 popsicle containers or a freezer tray. Freeze 4-6 hours or until hard.

Slender Strawberry Cooler

Use fresh or frozen strawberries

submitted by Bela Casson

Versatile is the word for this cool beverage—containing a variety of ingredients to add, substitute or omit.

Yield: 2 cups (500 ml)

1 cup (250 ml) fresh strawberries, washed and hulled
  or frozen unsweetened whole strawberries
1 cup (250 ml) skim milk
1/2 cup (125 ml) crushed ice (2-3 ice cubes)
1/4 cup (50 ml) plain low-fat yogurt
2 tablespoons (25 ml) sugar or equivalent artificial sweetener

In a blender, combine strawberries, skim milk, crushed ice, yogurt and sugar. Pour into glasses; garnish with a whole strawberry with stem.

Variation: Substitute vanilla yogurt for plain yogurt; add 1/2 teaspoon (2.5 ml) vanilla extract; add 1 teaspoon (5 ml) lemon juice; omit crushed ice; omit sugar if using frozen sweetened strawberries; substitute orange juice for skim milk.

Bela Casson is secretary of the Strawberry Growers Association of Nova Scotia. Her recipes have been featured at you-pick promotions, county fairs and other special occasions.

Strawberry Sorbet

Use fresh or frozen strawberries

submitted by Bela Casson

Made ahead of time, this recipe is great as a before-meal appetizer, light dessert or even between courses of a heavy meal.

Yield: 4 cups (1 L)

1	**cup (250 ml) sugar**
1	**cup (250 ml) boiling water**
1	**quart (1 L) fresh strawberries, washed and hulled**
	or 4 cups (1 L) frozen unsweetened strawberries
1	**cup (250 ml) orange juice**
1/4	**cup (50 ml) lemon juice**

In a bowl, dissolve sugar in water. Using a blender or food processor, combine strawberries, orange and lemon juices; blend until smooth. Add sugar mixture and blend. Transfer to a bowl or ice cube trays and freeze until hard (4-6 hours).

Break frozen mixture up in food processor until smooth as ice cream. Return to freezer and freeze until rehardened.

To serve, scoop into individual glass dishes. Garnish with mint leaf, whole strawberry with stem or green cherry.

Variation: Substitute artificial sweetener for sugar; sugared strawberries for fresh, reducing both sugar and water to 3/4 cup (175 ml).

The first Canadian cultivar testing program was begun at a federal research station in Kentville, Nova Scotia, in 1913. The well-known virus-free **strawberry** *plant propagation for Canada began in Nova Scotia in 1956. Today, three nurseries provide plants for local and export markets all over the world.*

Strawberry Growers of Nova Scotia
& Nova Scotia Federation of Agriculture

Sugar-Free Strawberry Pie

Use fresh strawberries

submitted by Elaine Maust

*This pie is especially good for the diabetic
and those desiring sugar-free desserts.*

Yield: 8 servings

1	pie shell (9 inches) (22.5 cm), baked
1	package (8 ounces) (22 g) sugar-free vanilla pudding (cooked type)
1	package (3 ounces) (85 g) sugar-free strawberry gelatin
2¹/₂	cups (625 ml) cold water
1	quart (1 L) fresh strawberries, washed and hulled

In a saucepan, mix dry pudding and gelatin with water. Stir over medium heat until mixture comes to a boil. Remove from heat. Cool until slightly thickened.

Arrange small whole or large sliced strawberries in pie shell. Pour cooled mixture over berries. Refrigerate until set. Serve with whipped topping, if desired. Add a whole strawberry with stem for garnish.

Duane and Elaine Maust have operated Sunshine Berry Farm near Meridian, Mississippi, since 1922. They currently grow eight acres of pick-your-own and ready-picked strawberries along with broccoli, sweet corn, and muscadine. The farm also boasts of a you-fish catfish lake.

Strawberry Bavarian Cake

Use fresh or frozen strawberries and jam

submitted by Susan Butler

We are always looking for ways to cut fat.
This cake may be low-fat, but it sure doesn't taste like it.

Yield: 8 servings

1 1/2 **pints (750 ml) fresh strawberries, washed, hulled, and chopped**
      or 2 cups (500 ml) chopped frozen unsweetened strawberries
  2/3 **cup (175 ml) powdered sugar**
  1/3 **cup (75 ml) plus 3/4 cup (200 ml) cold water**
   2   **envelopes unflavored gelatin**
1 1/2 **packages (3 ounces each) (85 g each) ladyfingers**
   2   **tablespoons (25 ml) strawberry jam**
   1   **teaspoon (5 ml) water**
  1/2 **cup (125 ml) light whipped topping, thawed**

Chop strawberries in blender or food processor. Do not puree.

In a saucepan, heat berries and sugar over medium heat until boiling, stirring occasionally. Reduce heat to low and simmer 5 minutes.

Pour mixture into medium mesh sieve over bowl. With a spoon, press berries against sieve to push through pulp and juice, equal to 2 cups (500 ml) puree.

In a saucepan, sprinkle gelatin over 1/3 cup (75 ml) cold water. Let stand 1 minute. Cook over low heat until gelatin dissolves. Stir into pureed fruit. Stir in 3/4 cup (200 ml) cold water. Refrigerate 1 1/4 hours, stirring occasionally until the consistency of unbeaten egg whites.

Separate ladyfingers into halves and line a 8 1/2x3-inch (21.25x7.5 cm) springform pan. Mix strawberry jam with 1 teaspoon (5 ml) water and spread over ladyfingers lining bottom of pan.

Fold whipped topping into thickened fruit mixture. Pour mixture into springform pan. Cover and refrigerate 2 hours until firm.

To serve, remove sides of springform pan and garnish with whipped topping, if desired.

Strawberry Main Dishes and Entrees

Doubtless God could have
made a better berry,
but doubtless God never did.

William Butler

Strawberry Crepes

Use fresh strawberries and strawberry jam

submitted by Bea Statz

*Crepes and strawberries are a great combination
for a Sunday family brunch—involve the kids in the preparation too.*

Yield: 24 (6-inch) (15 cm) crepes

1 1/2   **cups (375 ml) flour**
  1/8   **teaspoon (.5 ml) salt**
   3   **eggs**
1 1/2   **cups (375 ml) milk**
   2   **tablespoons (25 ml) butter or oil, melted and cooled**
   1   **cup (250 ml) or 1 small jar strawberry jam**
      Powdered sugar
   1   **quart (1 L) fresh strawberries, washed and hulled; whole or sliced**

In blender, mix together flour, salt, eggs, milk and butter; blend 1 minute. Scrape down sides of blender; blend 15 seconds more or until smooth.

Place crepe, omelet pan or heavy skillet over medium heat until hot. Brush with oil and remove pan from heat.

Pour in enough batter to cover bottom of pan. Swirl batter in pan to completely cover bottom in very thin layer. Return pan to heat and cook until crepe is set and edges are dry. Slide spatula under edge of crepe to loosen. Carefully lift and gently turn crepe to brown other side; remove pan from heat. Shake pan to loosen crepe; slide onto oiled waxed paper.

Lightly spread half of each crepe with about 1 tablespoon (15 ml) jam. Fold in half, then in quarters. Sprinkle with powdered sugar. Garnish with fresh strawberries.

Variation: Add whipped topping or yogurt to jam; substitute pancakes for crepes; 1 package ready-to-use frozen crepes.

 Strawberries *are found all over the world... "from Kashmir to Kamchatka to Spain, Oregon and Hudson's Bay—in deep valleys or alpine heights."*

The Strawberry Connection

Strawberry French Toast with Pocket

Use frozen strawberries

submitted by Carolyn Beinlich

*There is a pocket of creamy filling in every bite of
French toast smothered in a delicious strawberry sauce.*

Yield: 10 servings

Sauce:
- 1 package (10 ounces) (283 g) frozen sweetened strawberries, thawed
- 1/2 cup (125 ml) sugar
- 2 tablespoons (25 ml) cornstarch
- 1/2 cup (125 ml) sliced almonds, toasted

Filling:
- 1 package (8 ounces) (227 g) cream cheese
- 2 tablespoons (25 ml) sugar
- 1 teaspoon (5 ml) vanilla extract

French toast:
- Margarine for browning
- 1 loaf (15x5 inches) (37.5x12.5 cm) Italian bread, unsliced
- 6 eggs, beaten
- 1/3 cup (75 ml) milk
- 1/4 teaspoon (1 ml) nutmeg

To make sauce, drain strawberries, reserving juice. Add water to reserved juice to make 1 cup (250 ml). Combine sugar and cornstarch in saucepan; gradually add reserved juice and stir. Cook over medium heat, stirring constantly until mixture is clear and thickened. Stir in strawberries and toasted almonds. Remove from heat; serve warm.

To make filling, in a mixer or food processor blend cream cheese, sugar and vanilla. Mix until well blended.

To make French toast, grease and heat griddle. Cut bread into 1 1/2-inch-thick (3.75 cm) slices. Cut slit through crust of each slice to form pocket. Fill each pocket with 1 rounded measuring tablespoon (15 ml) cream cheese mixture.

In a small bowl, beat eggs and milk until well blended. Dip each slice into egg mixture. Grill on both sides until golden brown. Serve with warm strawberry sauce.

Fiesta Strawberry Bowl

Use fresh strawberries

submitted by Linda Yoder

Yield: 8 servings

Fruit mixture:
- 3 cups (750 ml) washed, hulled and halved fresh strawberries
- 2 cups (500 ml) fresh pineapple chunks
- 1 cup (250 ml) fresh peeled orange slices
- 1 teaspoon (5 ml) lime zest
- 2 tablespoons (25 ml) honey (optional)

Topping:
- 2 cups (500 ml) plain yogurt
- 2 tablespoons (25 ml) honey or to taste
- 1 teaspoon (5 ml) cinnamon or to taste

Sesame Puff Bowl:
- 1 tablespoon (15 ml) sesame seed, divided
- 1/2 cup (125 ml) water
- 1/4 cup (50 ml) butter, cut up
- 1/2 cup (125 ml) flour
- 2 large eggs

To make fruit mixture, combine strawberries, pineapple chunks, orange slices, lime zest and honey; marinate while making Sesame Puff Bowl.

To make topping, mix yogurt, honey and cinnamon. Refrigerate until ready to serve.

Preheat oven to 450 degrees F (240 C). Spray a 9-inch (22.5 cm) pie pan with nonfat cooking spray and sprinkle with 1 1/2 teaspoons (7 ml) of the sesame seed.

In a saucepan, heat water and butter over high heat until the water boils and butter melts. Reduce heat and add flour all at once, stirring quickly with a wooden spoon until dough forms a ball that leaves the sides of the pan. Remove from heat; cool 5 minutes.

Add eggs, one at a time, beating after each addition until mixture is smooth. Spoon mixture into pie pan and spread to cover bottom. Sprinkle with the remaining 1 1/2 (7 ml) teaspoons sesame seed.

Bake 15 minutes until sides puff up and form a bowl. Reduce temperature to 350 degrees F (180 C) and bake 10 minutes more until golden brown. Cool slightly.

Fill warm puff with marinated fruit and add dollops of yogurt mixture or serve on the side.

 Del and Linda Yoder's family of Morgantown, West Virginia, began a pick-your-own strawberry operation in 1978. Much to their surprise, evidence of a former 60-year-old commercial strawberry farm was discovered.

Strawberry Salsa for Meat, Poultry, Fish or Chips

Use fresh strawberries

submitted by Bela Casson

*Strawberry Salsa is great as a side dish or as a snack with chips.
You will find this recipe very versatile!*

Yield: 2 cups (500 ml)

2   **cups (500 ml) washed, hulled and diced fresh strawberries**
1/2 **cup (125 ml) diced green pepper**
2   **green onions, diced**
1   **teaspoon (5 ml) salt**
1   **teaspoon (5 ml) dried basil**
1   **tablespoon (15 ml) lemon juice**
1   **tablespoon (15 ml) vegetable oil**
2   **tablespoons (25 ml) honey**
Dash of cayenne pepper

In a medium bowl combine strawberries, green pepper, onions, salt, basil, lemon juice, oil, honey and cayenne pepper; mix well. Cover and refrigerate 2 hours to blend flavors.

Serve salsa on the side as an accompaniment to meat, poultry or fish.

Variation: Substitute cilantro for basil.

The first record of **strawberry** *export from Nova Scotia occurred in Yarmouth when there was a shipment to Boston in 1868. Five buckets of the "Wilson" variety were purchased for 5 shillings (about 10 cents per quart).*

Strawberry Growers of Nova Scotia
& Nova Scotia Federation of Agriculture

Strawberry Bean Bake

Use fresh strawberries and jam

submitted by Bea Statz

This unusual baked bean dish was entered in the 1995 Wisconsin Berry Recipe Contest by Leora M. McCarthy of Fox Lake, Wisconsin.

Yield: 12 servings

2   cans (1 pound each) (454 g each) baked beans
(with bacon and brown sugar base)
1/2   cup (125 ml) strawberry jam
2   teaspoons (10 ml) fruit pectin

Topping:
1/3   cup (75 ml) flour
1/2   cup (125 ml) brown sugar
2   tablespoons (25 ml) butter
1/8   teaspoon (.5 ml) salt
3/4   cup (200 ml) washed, hulled and sliced fresh strawberries
1   can (2.8 ounces) (79 g) french fried onions
1/4   teaspoon (1 ml) grated nutmeg

Preheat oven to 350 degrees F (180 C).

Place beans in a 1 1/2-quart (1.5 L) baking dish. Stir in jam and fruit pectin; set aside.

To make topping, in a small bowl, combine flour, brown sugar, butter and salt. Using a pastry blender or fork, cut butter into mixture. Toss lightly with sliced strawberries and spoon over beans.

Sprinkle top with french fried onions and nutmeg. Bake 30-45 minutes.

Serve with lettuce salad and ham.

 Mexico exports over half of their crop of **strawberries**, *mainly to the United States.*

The Strawberry Connection

Strawberry Barbecued Chicken

Use fresh strawberries

submitted by Helen Thoren Kent

Yield: 6 servings

6   boneless chicken breasts
Worcestershire sauce
Salt and pepper
1   jar (12 ounces) (340 g) red currant jelly
1   pint (500 ml) fresh strawberries, washed, hulled and sliced

Cook chicken breasts over an outdoor grill. Season to taste with Worcestershire, salt and pepper.

In a small saucepan, heat currant jelly slowly until warm and syrupy. Stir in strawberries and heat until berries are just warmed through.

Serve warm over chicken breasts.

Strawberry Wings

Use strawberry Jam

submitted by Linda Yoder

Yield: 8 servings

2¹/₂   pounds (1200 g) chicken wings
2   cloves garlic, crushed
1   tablespoon (15 ml) grated fresh gingerroot
3   tablespoons (45 ml) fresh lemon juice
¹/₃   cup (75 ml) soy sauce
Crushed hot pepper flakes to taste (optional)
³/₄   cup (200 ml) strawberry jam or preserves

Wash and trim chicken wings; Place in 9x13-inch (22.5x30.5 cm) baking dish to marinate. Combine garlic, ginger, lemon juice, soy sauce and hot pepper flakes; pour over wings and marinate 1 hour. Drain and reserve marinade. Combine reserved marinade with strawberry jam.

Arrange chicken wings, upside down, in baking dish; bake 20 minutes. Turn over wings and coat with reserved marinade-jam mixture. Bake 20 minutes more or until wings are tender and brown.

Before serving, garnish with strawberry fans and scallions, if desired.

Chicken Salad with Pita Bread

Use fresh strawberries

submitted by Elaine Maust

This recipe is low-fat and low-calorie—a real healthy luncheon idea.

Yield: 4 servings

  2   **cups (500 ml) cooked, cooled and diced chicken**
  1   **cup (250 ml) shredded lettuce**
3/4 **cup (200 ml) halved white grapes**
  1   **cup (250 ml) washed, hulled and halved fresh strawberries**
  1   **cup (250 ml) alfalfa sprouts**
  1   **cup (250 ml) light dressing of your choice**
    Pita bread or spinach

In large bowl combine chicken, lettuce, grapes, strawberries and sprouts. Toss gently with dressing. Stuff into pita bread or serve on a bed of spinach.

Variation: Substitute turkey for chicken.

 The **strawberry** *belongs to the genus* Fragaria, *a member of the rose family,* Rosaceae, *which includes roses, apples and plums. The genus is so widely distributed that its fruit are gathered by the Laplanders in the Arctic and by the people in India.*

The Compleat Strawberry

Honey Chicken with Strawberries

Use fresh strawberries

submitted by Bea Statz

*This recipe is wonderful with tossed salad and rice.
It is adapted from a recipe in* It's the Berries *by Anton & Dooley.*

Yield: 8 servings

1   **large chicken, quartered or 6 breasts**
Salt and pepper to taste
2   **tablespoons (25 ml) butter or margarine**
2   **tablespoons (25 ml) vegetable oil**
4   **shallots or minced green onions**
1/2   **cup (125 ml) Strawberry Vinegar (see page 18)**
1/3   **cup (75 ml) honey**
1 1/2   **cups (375 ml) washed, hulled and halved fresh strawberries**

Rinse and dry chicken pieces, removing skin, if desired. Season with salt and pepper.

Preheat large skillet on medium heat; add butter and oil. Brown chicken and remove from skillet.

Saute shallots in skillet until translucent, adding oil if necessary. Add Strawberry Vinegar and honey to skillet; simmer 1 minute. Return chicken to skillet; cook partially covered for approximately 20-30 minutes. Baste often to glaze.

Toss in strawberries just before serving.

The parts of the **strawberry**, *which we think of as the fruit, is in fact a false fruit, a receptacle at the end of the flower stalk. The tiny achenes (seeds) on its surface are the true fruits of the plant. They are neither surrounded by juice nor covered with a skin; they sit naked on the outside of the fruit.*

The Compleat Strawberry

Strawberry Chicken Salad

Use fresh strawberries

submitted by Virginia Radewald

Serve this salad with a muffin and dessert at a luncheon, party or shower.

Yield: 6 servings

2 1/2   **cups (625 ml) cooked, cooled and cubed chicken breasts**
  2   **cups (500 ml) washed, hulled and halved fresh strawberries; reserve 6 whole perfect strawberries with hull for garnish**
  1   **cup (250 ml) seedless grapes**
  1   **cup (250 ml) pineapple chunks**
  1   **cup (250 ml) diced celery**
  1/2   **cup (125 ml) coarsely chopped pecans**
  1/2   **cup (125 ml) sweet-sour or poppy seed bottled salad dressing**
  3   **cups (750 ml) spinach, washed and torn in bite-size pieces**
  3   **cups (750 ml) romaine lettuce, washed and torn in bite-size pieces**

Combine chicken, strawberries, grapes, pineapple, celery and pecans. Toss gently with salad dressing.

Mix spinach and lettuce together; arrange on individual salad plates. Spoon chicken mixture onto salad greens. Garnish each top with whole strawberry.

Note: Salad ingredients can also be tossed with spinach and lettuce greens and served in a bowl.

Variation: Substitute turkey for chicken.

 Edwin and Virginia Radewald own a four-generation 100-year-old farm six miles north of Niles, Michigan. The strawberry operation grew from seven acres in 1957 to a peak of 60 acres and included pick-your-own strawberries plus direct and wholesale marketing. The last crop of berries was produced in 1995. Today, fruit trees, vegetables, a cold storage facility and packing house complete the operation.

Strawberry Turkey Salad with Mint Dressing

Use fresh strawberries

submitted by Carolyn Beinlich

This magnificent main dish salad was adapted from a recipe in Country Woman *magazine.*

Yield: 4 servings

1   **pint (500 ml) fresh strawberries, washed, hulled and sliced**
1   **teaspoon (5 ml) sugar**
2   **teaspoons (10 ml) red wine vinegar**
3   **tablespoons (45 ml) vegetable oil**
1   **teaspoon (5 ml) dried mint**
1/2 **teaspoon (2 ml) salt**
2   **cups (500 ml) cooked, cooled and cubed turkey breast**
1/4 **cup (50 ml) sliced toasted almonds**
Lettuce
Whole mint leaves for garnish

In medium bowl, toss sliced strawberries with sugar and vinegar. Let stand 30 minutes.

In large bowl, mix oil, mint and salt until blended. Add turkey and stir to coat. Refrigerate.

Toast almonds in 350-degree F (180 C) oven 5-10 minutes.

Just before serving, gently stir in strawberry mixture and toasted almonds with turkey mixture. Serve on lettuce and garnish with mint leaves.

Variation: Substitute chicken for turkey.

 The Spanish explorers found **strawberries** *on the sand dunes of California in December 1602.*

Shrimp Strawberry Salad

Use fresh strawberries

submitted by Bea Statz

There are unique and complimentary flavors in this shrimp and strawberry entree. It is adapted from a recipe in **Simply Strawberries** *by Sara Pitzer.*

Yield: 6 servings

1 1/2   pounds (720 g) cooked, shelled and deveined shrimp (fresh or frozen); reserve a few shrimp for garnish
1   cup (125 ml) diced celery
4   hard-boiled eggs, chopped
3   tablespoons (45 ml) chopped green pepper

Dressing:
1   teaspoon (5 ml) dry mustard
2   tablespoons (25 ml) chopped fresh parsley
1   tablespoon (15 ml) lemon juice
3/4   cup (200 ml) mayonnaise
1/4   cup (50 ml) sour cream
2   cups (500 ml) fresh strawberries, washed and hulled; reserve a few strawberries for garnish
Lettuce greens

In a large bowl, combine shrimp, celery, eggs and green pepper.

To make dressing, combine mustard, parsley, lemon juice, mayonnaise and sour cream. Beat with a fork or mix in blender or food processor. Pour over shrimp mixture; toss lightly. Add strawberries; stir gently to mix. Refrigerate 30 minutes.

Serve on lettuce greens. Garnish with reserved shrimp and strawberries.

 You have to eat eight cups of sweet cherries to get the same amount of Vitamin C found in a single cup of **strawberries**. *Both cherries and* **strawberries** *are a low-calorie, low-fat snack when you want a delicious, guilt-free dessert. But to get your daily (60 milligram) dose of Vitamin C, remember it's the berries!*

American Health magazine

Stuffed Pork Chops with Strawberries

Use fresh strawberries

submitted by Bea Statz

This is a delightful main dish blending pork and strawberry flavors. It is adapted from a recipe in Old-Fashioned Strawberry Recipes *by Bear Wallow Books.*

Yield: 4 servings

4   **thick pork chops, with pocket cut for stuffing**
3   **cups (750 ml) fresh strawberries, washed and hulled; divided**

Stuffing:
1/2   **cup (125 ml) dry bread crumbs or unseasoned dressing cubes**
1   **tablespoon (15 ml) hot water**
2   **tablespoons (25 ml) butter or margarine, melted**
1   **tablespoon (15 ml) diced celery**
1   **tablespoon (15 ml) minced onion**
1   **tablespoon (15 ml) minced green pepper**
1/2   **teaspoon (2 ml) salt**
1/2   **teaspoon (2 ml) pepper**
1/2   **teaspoon (2 ml) sage**
1   **tablespoon (15 ml) cooking oil**
4   **tablespoons (60 ml) brown sugar**
1/4   **cup (50 ml) soy sauce**

Preheat oven to 375 degrees F (190 C). Spray 9x13-inch (22.5x30.5 cm) baking pan with nonfat cooking spray.

If strawberries are large, halve or slice 1 cup (250 ml). Combine strawberries, bread crumbs, water, butter, celery, onion, green pepper, salt, pepper and sage; mix well. Stuff pork chops with mixture and close with toothpicks. Rub pork chops with cooking oil and season with salt and pepper.

Place pork chops in prepared pan in oven, uncovered, for 15 minutes or until chops are browned.

Mash remaining 3 cups strawberries in saucepan. Add brown sugar and soy sauce. Heat until sugar is dissolved. Baste pork chops with half of the mixture. Cover pan with foil and continue baking 30 minutes. Baste with remaining strawberry mixture and bake 15 minutes longer or until chops are tender; reduce heat if necessary. Remove toothpicks before serving.

Baked Ham with Strawberry Glaze

Use frozen strawberries

submitted by Marilyn Gjevre

A spicy sweet-sour glaze for ham, prepared with strawberries from the freezer, is a delightful way to treat family and friends.

Yield: 1 1/2 cups (375 ml)

1   **thick slice ham**
Margarine for browning
1   **tablespoon (15 ml) dry mustard**
4   **tablespoons (60 ml) brown sugar**
1/3   **cup (75 ml) white vinegar**
1   **package (10 ounces) (283 g) frozen sweetened strawberries, thawed and undrained**
2   **sticks cinnamon**
2   **whole cloves**
1/2   **teaspoon (2 ml) ginger**
Dash of nutmeg

Preheat oven to 325 degrees F (140 C).

Score ham and cook in hot skillet; turn until lightly browned on both sides. Place a 9x13-inch (22.5x30.5 cm) baking dish; set aside.

Combine mustard, brown sugar and vinegar; spread over ham.

In saucepan, combine strawberries, cinnamon, cloves and ginger; simmer 10-15 minutes. Remove cinnamon sticks and cloves; pour over ham and bake covered 1 1/2 hours. Serve hot.

*The main positive attribute of **strawberries** is their taste: "when they're perfect, there is nothing quite like them." Part of this taste advantage is tied to appearance. Anticipation of eating **strawberries** is a strong part of the consumption experience.*

"Strawberries: Consumer Attitudes and Usage,"
Queensland Fruit & Vegetable Growers

Strawberry Beverages and Punches

Curly locks, curly locks,
wilt thou be mine? Thou shalt
not wash dishes, nor yet feed
the swine, but sit on a fine
cushion and sew a fine seam
and feed upon **strawberries**,
sugar and cream.

Traditional

Strawberry Milk Shake

Use fresh or frozen strawberries

submitted by Queensland Fruit & Vegetable Growers

Home economist Abigail Walsh includes this nutritious drink recipe in Queensland, Australia's, The Market Link.

Serves: 4

1	**pint (500 ml) fresh strawberries, washed and hulled**
2 3/4	**cups (700 ml) milk**
3/4	**cup (200 ml) vanilla yogurt**
2	**tablespoons (25 ml) honey**
1	**teaspoon (5 ml) lemon juice**
	Whole strawberry or mint leaf for garnish

Using a blender, combine strawberries, milk, yogurt and honey. Blend until smooth. Add lemon juice and mix until well blended.

To serve, pour into glasses half filled with ice. Garnish each glass with a whole strawberry or mint leaf.

Variation: Use 1 to 2 cups (250-500 ml) frozen unsweetened strawberries or 1 to 2 packages (10 ounces each) (283 g each) frozen sweetened strawberries, omitting honey.

Strawberry *shortcakes was actually one of the earlier culinary concoctions and probably derived from a Native American dish. The first* **strawberry** *shortcake wa a slightly sweetened biscuit mounded over with fresh wild* **strawberries** *and smothered with ladles of fresh warm cream.*

Old-Fashioned Strawberry Recipes With Historic Notes

Strawberry Fruit Punch

Use fresh or frozen strawberries

submitted by Elaine Maust

This colorful and refreshing punch is perfect for those special occasions such as parties or weddings. Try substituting cranberry for a festive touch at Christmas.

Yield: 20 servings

1 pint (500 ml) fresh strawberries, washed and hulled
1 can (46 ounces) (1360 g) pineapple juice
1 can (12 ounces) (355 g) frozen pink lemonade concentrate
1 can (12 ounces) (355 g) frozen orange juice concentrate
2 cans (12 ounces each) (355 g) water
2 quarts (2 L) ginger ale, chilled

Puree strawberries in a blender or food processor until liquid.

In a large container or punch bowl, add strawberries, pineapple juice, lemonade concentrate, orange juice concentrate and water. Mix well. Add ginger ale and ice cubes or ice ring just before serving.

Variation: Substituting cranberry juice for pineapple juice adds a tangy flavor. Use 2 cups (500 ml) frozen unsweetened strawberries or 2 packages (10 ounces each) (283 g each) frozen sweetened strawberries, partially thawed and pureed.

California and Florida account for 90 percent of United States table **strawberries.**

Strawberry Sparkle

Use fresh or frozen strawberries

submitted by Bela Casson

A naturally refreshing drink that's great for entertaining anytime of the year and appeals to all ages.

Yield: 4 quarts (4 L)

1   **quart (1 L) fresh strawberries, washed and hulled**
1   **cup (250 ml) sugar**
1/2 **cup (125 ml) water**
1/4 **cup (50 ml) lemon juice**
4   **cups (1 L) unsweetened apple juice**
8   **cups (2 L) raspberry ginger ale or fruit-flavored mineral water**

Combine strawberries, sugar, water and lemon juice in blender; blend on high until smooth.

Pour mixture into punch bowl; add apple juice and ginger ale. Stir and serve.

Variation: Use 4 cups (1 L) frozen unsweetened strawberries if fresh are not available.

 Nova Scotians love their favorite fruit. They eat the most **strawberries** *of any province in Canada, about 4 quarts per capita.*

Strawberry Growers of Nova Scotia
& Nova Scotia Federation of Agriculture

Strawberry Lemonade Concentrate

Use fresh strawberries

submitted by Bea Statz

This unusual concentrate recipe was discovered in **Country Woman** *magazine. It's a refreshing beverage to be enjoyed during and long after the strawberry season is over—a great way to preserve an abundance of strawberries.*

Yield: 6 quarts (6 L)

4 quarts (4 L) fresh strawberries, washed and hulled
4 cups (1 L) fresh lemon juice (about 16 lemons)
3 quarts (3 L) water
6 cups (1.5 L) sugar

Puree strawberries in a blender or food processor.

Place strawberries in a large kettle; add lemon juice, water and sugar. Heat to 165 degrees F (70 C) over medium heat, stirring occasionally. (Do not boil.) Remove from heat; skim off foam.

Pour into hot sterilized jars leaving 1/4 inch (1 cm) headspace. Adjust caps and lids. Using the water-bath method, place jars in canner or large saucepan. Add enough water to cover tops of jars by 1-2 inches (2.5-5.0 cm). Cover canner and bring to a boil; boil 15 minutes. Remove jars from canner. Let stand to seal and cool.

To serve, mix about one-third (75 ml) to two-thirds cup (175 ml) with soda or ginger ale and pour over ice cubes. Refrigerate after opening. Use within 1 year.

 Prior to 1766 no Europeans seemed aware of the importance of sex in **strawberries***. Female plants must be fertilized with pollen from male plants.*

The Strawberry Connection

Strawberry Daiquiri

Use fresh or frozen strawberries

submitted by Bea Statz

The daiquiri is a classic strawberry cocktail.

Yield: 1 or 6 daiquiris

1 serving:
- 4-6 fresh whole strawberries, washed and hulled or frozen unsweetened whole strawberries
- 1 jigger (1 1/2 ounces) (45 g) light rum
- 1 tablespoon (15 ml) fresh lime juice
- 1 teaspoon (5 ml) sugar
- 4-5 ice cubes

6 servings:
- 1 1/2 cups (375 ml) fresh whole strawberries, washed and hulled or frozen unsweetened whole strawberries
- 3/4 cup (200 ml) light rum
- Juice of 1 lime
- 2 tablespoons (25 ml) sugar
- 1 1/2 trays ice cubes

Combine strawberries, rum, lime juice and sugar in blender. Process until strawberries are pureed. Add ice cubes and blend until mixture is slushy.

Serve in champagne glasses. Garnish using a fresh whole strawberry with hull.

 *Enhancing your aerobic workout with a whiff of **strawberry** scent helped exercisers burn more calories in a three-minute workout than either bad odors or the absence of scents.*

Fitness magazine

Strawberry Margarita

Use fresh or frozen strawberries

submitted by Bea Statz

This is definitely a party drink and is great with Mexican food.

Yield: 1 or 6 margaritas

1 serving:

4-6	fresh whole strawberries, washed and hulled or frozen unsweetened whole strawberries
1 1/2	ounces (45 g) tequila
1/2	ounce (15 g) triple sec
1	ounce (30 g) lemon or lime juice
4-5	ice cubes

Salt to prepare glasses
Rind of lemon or lime to prepare glasses

6 servings:

1-1 1/2	cups (250-375 ml) fresh strawberries, washed and hulled or frozen unsweetened whole strawberries
9	ounces (270 g) tequila
3	ounces (85 g) triple sec

Juice of 1 lemon or lime, reserve rind
1 1/2 trays ice cubes
Salt to prepare glasses

Combine strawberries, tequila, triple sec and lemon juice in blender. Process until strawberries are pureed. Add ice cubes and blend until mixture is shaved, but not slushy.

Rub rim of champagne glass with rind of lemon or lime. Dip rim in salt. Pour in blended mixture. Garnish using a fresh whole strawberry with hull.

 On the average, there are 400 tiny seeds on every **strawberry**.

California Strawberry Advisory Board

Strawberry-Wine Punch

Use fresh or frozen strawberries

submitted by Susan Butler

Make this punch for parties, special occasions or the holidays!

Yield: 10 servings

1   **can (6 ounces) (177 g) frozen lemonade concentrate**
2   **cups (500 ml) washed, hulled and halved fresh strawberries
     or 1 package (10 ounces) (283 g) frozen sweetened strawberries**
1   **bottle (25.3 fluid ounces) (750 ml) white or blush wine**
1   **bottle (1 L) ginger ale**
1   **lemon, sliced**
Ice (frozen in Bundt cake pan or ring jello mold)

In a blender, mix lemonade concentrate, strawberries and wine until pureed. Pour mixture into punch bowl. Add ginger ale, lemon and ice just before serving.

Variation: Non-alcoholic white wine and frozen unsweetened strawberries will make the punch lighter.

Butler Orchards in Germantown, Maryland, was started in 1950 by NASGA founder George Butler and his wife, Shirley. The family-operated premiere pick-your-own farm now grows 20 acres of strawberries along with other fresh fruits, vegetables, flowers and Christmas trees for public harvest.

Strawberry Pies

Strawberry leaves have a most cordial smell, next in sweetness to the musk rose and violet.

Francis Bacon

Quick and Easy Strawberry Pie

Use fresh strawberries

submitted by Debbie Lineberger

A customer shared this recipe for "quick and easy" strawberry pie. It can easily be doubled and is great for large gatherings or for taking to a potluck.

Yield: 1 pie (9 inches) (22.5 cm)

1   **quart (1 L) fresh strawberries, washed, hulled and sliced**
5   **slices white or whole wheat sandwich bread**
1/2   **cup (125 ml) butter or margarine**
3   **tablespoons (45 ml) flour**
1   **egg**
1   **cup (250 ml) sugar**
1   **teaspoon (5 ml) vanilla extract**

Preheat oven to 350 degrees F (180 C). Spray an 8-inch (20 cm) square baking pan with nonstick cooking spray. Spread sliced strawberries in prepared pan.

Cut bread into narrow strips and lay crisscross over berries. Melt butter in large microwave-safe bowl. Add flour, egg, sugar and vanilla; mix with whisk. Pour mixture over bread, coating evenly. Bake 30 minutes or until lightly browned. Remove from heat; cool.

Note: This recipe can easily be doubled. Use a 9x13x2-inch (22.5x30.5x5 cm) pan and bake 40 minutes.

Ervin and Debbie Lineberger of Kings Mountain, North Carolina, farm 100 acres of pick-your-own fruits and vegetables, with strawberries being the anchor crop. Located in the Southern Piedmont Region near the South Carolina border, they also make jams and jellies to sell in holiday gift boxes and baskets.

Sunshine Strawberry Pie

Use fresh or frozen strawberries

submitted by Elaine Maust

This easy pie is the most requested recipe at Sunshine Berry Farm.

Yield: 1 pie (9 inches) (22.5 cm)

1	**pie shell (9 inches) (22.5 cm) baked**
1 1/2	**cups (375 ml) water**
3/4	**cup (200 ml) sugar**
2	**tablespoons (25 ml) cornstarch**
3	**tablespoons (45 ml) strawberry gelatin**
1	**teaspoon (5 ml) lemon juice**
1	**quart (1 L) fresh strawberries, washed and hulled**
	or frozen unsweetened whole strawberries, partially thawed

In saucepan, mix water, sugar and cornstarch. Heat to a boil; boil for 1 minute. Remove from heat; add gelatin and lemon juice, stirring until dissolved. Cool slightly, then add strawberries. Pour into cooled crust. Refrigerate until set.

Serve with whipped cream or whipped topping.

Variation: When using frozen unsweetened whole strawberries, increase cornstarch to 3 tablespoons (45 ml) and gelatin to 4 1/2 tablespoons (65 ml).

Strawberries *were loved by royalty. Princess Mary, King Henry's daughter, liked them so much she was usually greeted with a basket of fresh red* **strawberries** *when she visited the Court.*

Home Cooking magazine

Strawberry Parfait Pie

Use fresh or frozen strawberries

submitted by Virginia Radewald

Because it can be made ahead of time and utilizes either fresh or frozen strawberries in the off-season, this recipe is a family favorite.

Yield: 1 pie (9 inches) (22.5 cm)

1   **pie shell (9 inches) (22.5 cm) baked
    or prepared graham cracker crust**
1   **cup (250 ml) boiling water**
1   **package (3 ounces) (85 g) strawberry gelatin**
2   **cups (500 ml) vanilla or strawberry ice cream**
1   **cup (250 ml) washed, hulled and sliced fresh strawberries
    or frozen unsweetened sliced strawberries**

In a large bowl, add boiling water to gelatin and stir until dissolved. Spoon ice cream into gelatin and stir until melted. Chill until mixture starts to thicken. Add strawberries. Pour into prepared pie shell. Refrigerate until firm.

Garnish with whipped topping and/or fresh whole strawberries.

Variation: Frozen sweetened strawberries, thawed and drained, can also be used. The drained strawberry juice may be added to the water to make 1 cup (250 ml) of liquid.

 Shaped like a heart and colored bright red, the **strawberry** *is a natural symbol of love. So it's no wonder that the* **strawberry** *was also a symbol for Venus, the goddess of love.*

Home Cooking magazine

Deluxe Fresh Strawberry Pie

Use fresh strawberries

submitted by Ann Epp

This scrumptious pie should be served to family and guests at least once each strawberry season.

Yield: 1 pie (9 inches) (22.5 cm)

1   **pie shell (9 inches) (22.5 cm) baked or crumb crust of your choice**
1   **package (3 ounces) (85 g) cream cheese, softened**
3   **tablespoons (45 ml) milk**

Glaze:
2   **cups (500 ml) fresh whole strawberries, washed and hulled; divided**
1/2   **cup (125 ml) sugar**
1/8   **teaspoon (.5 ml) salt**
1/2   **cup (125 ml) water, divided**
1/4   **cup (50 ml) cornstarch**
1   **tablespoon (15 ml) lemon juice**

In a small bowl, beat cream cheese and milk together. Spread over bottom of prepared crust; set aside.

To make glaze, place 1 cup (250 ml) of the strawberries in saucepan. Chop strawberries into a sauce. Add sugar, salt and 1/4 cup (50 ml) of the water. Bring to a boil over medium heat and cook 3 minutes, stirring constantly. Combine cornstarch and the remaining 1/4 cup (50 ml) water; add to glaze mixture and cook until thick, stirring constantly. Add lemon juice and remove from heat. Cool.

Spread one half of cooled glaze over cream cheese layer. Place the remaining 1 cup (250 ml) whole berries, with top end down, on top of glaze. Spoon remaining glaze over berries. Refrigerate until ready to serve.

Garnish with whipped topping, if desired.

 Ben and Ann Epp have been in the berry business for 23 years and started growing strawberries in Saskatoon in 1981. The farm, which also grows raspberries and saskatoons, is located four miles southwest of Saskatoon, Saskatechewan.

Fresh Strawberry Cream Cheese Pie

Use fresh strawberries

submitted by Marion Borchert

Betty Crocker gets credit for this easy, delicious and pretty pie.

Yield: 1 pie (9 inches) (22.5 cm)

1	**prepared graham cracker crust (9 inches) (22.5 cm)**
1 1/2	**quarts (6 cups) (1.5 L) fresh strawberries, washed and hulled; divided**
1	**cup (125 ml) sugar**
3	**tablespoons (45 ml) cornstarch**
1/2	**cup (125 ml) water**
1	**package (3 ounces) (85 g) cream cheese, softened**

Whipped cream or whipped topping for garnish

Mash 2 cups (500 ml) of the strawberries to make 1 cup (250 ml) sauce. In saucepan, stir together sugar and cornstarch. Gradually stir in water and strawberries. Heat over medium heat until mixture thickens, stirring constantly. Bring to a boil; boil 1 minute. Remove from heat to cool.

Spread softened cream cheese over graham cracker crust.

Use enough of remaining 4 cups (1 L) strawberries (either whole or halved) to fill pie shell, reserving whole strawberries for garnish. Pour cooled mixture over berries. Refrigerate until set.

Before serving, top with whipped topping and decorate edge with reserved whole berries.

In business for 75 years, Borchert Orchards is located in New York's Hudson River Valley. The four acres of strawberries are mostly sold prepicked along with a variety of fruits, berries and vegetables.

Ritzy Strawberry Pie

Use fresh strawberries

submitted by Marilyn Gjevre

First magazine carried this recipe which is served at Steinbeck House in Salinas, California. The particularly different and delicious crust complements the strawberry and cream filling. Wait for the ohs and aahs!

Yield: 1 pie (9 inches) (22.5 cm)

Crust:
- 1 **cup (250 ml) ground Ritz Crackers (25-30 crackers)**
- 1 **cup (250 ml) coarsely chopped walnuts**
- 1/2 **teaspoon (2 ml) baking powder**
- 1 **cup (250 ml) sugar**
- 4 **egg whites**
- 1 **teaspoon (5 ml) vanilla extract**

Filling:
- 1 **cup (250 ml) heavy whipping cream**
- 2 **tablespoons (25 ml) sugar**
- 1/2 **teaspoon (2 ml) vanilla extract**
- 1-2 **pints (500 ml-1 L) washed, hulled and sliced fresh strawberries**

Preheat oven to 325 degrees F (160 C). Spray a 9-inch (22.5 cm) pie plate with nonstick cooking spray; set aside.

To make crust, combine crackers and walnuts in a bowl; set aside.

In a separate bowl, combine baking powder and sugar; set aside.

In an electric mixing bowl, beat egg whites until stiff peaks form. Gradually add sugar mixture, beating constantly. Continue beating until whites hold stiff peaks. Add vanilla and continue beating. Fold cracker mixture into beaten egg whites. Spread into prepared pan. Bake 40-50 minutes until golden brown. Remove from oven and cool.

To make filling, beat whipping cream, sugar and vanilla until cream holds stiff peaks. Spread over cooled crust. Garnish top with sliced strawberries. Refrigerate until served.

 California harvests enough **strawberries** *if laid berry to berry would wrap around the world nearly 15 times.*

California Strawberry Advisory Board

Strawberry Pizza Pie

Use fresh strawberries

submitted by Doris Stutzman

*This is an eye-catching addition for any occasion,
and it's great as a snack, hors d'oeuvre or dessert.*

Yield: 1 pie (9 inches) (22.5 cm)

Crust:
- 2   **cups (500 ml) flour**
- 1   **cup (250 ml) butter
  or margarine**
- 4   **tablespoons (60 ml)
  powdered sugar**

Filling:
- 2   **packages (3 ounces each) (85 g each)
  cream cheese, softened**
- 1   **container (8 ounces) (226 g)
  whipped topping**
- 1   **cup (250 ml) powdered sugar**
- 1   **pint to 1 quart (500 ml-1 L)
  fresh strawberries, washed,
  hulled and halved**
- 1   **package (16 ounces) (454 g)
  prepared strawberry glaze, divided**

Preheat oven to 400 degrees F (200 C). Spray 13x9x2-inch (22.5x30.5x5 cm) glass baking dish with nonstick cooking spray; set aside.

In a medium bowl using a fork or pastry blender, cut flour, butter and powdered sugar into coarse crumbs. Place in prepared dish. Bake 8-10 minutes or until lightly browned. Remove from oven and cool.

In a large bowl, mix together cream cheese, whipped topping and powdered sugar. Spread over prepared crust and refrigerate until ready to use.

Just before serving, spread one half of the strawberry glaze over cream cheese layer. Arrange halved strawberries decoratively·on top. Cover berries with remaining glaze. Refrigerate until ready to serve

Variation: Substitute blueberries, kiwi, peaches, pears, etc. for strawberries.

 *Early in the 1300s, monks in western Europe used wild **strawberries** in their "illuminated" manuscripts. The Madonna was often shown among the **strawberries**.*

The Strawberry Connection

Heavenly Chocolate Berry Pie

Use fresh strawberries

submitted by Susan Butler

Assemble crust and filling a day ahead and top it off with strawberries and chocolate drizzle just before serving. This creation will become a family favorite.

Yield: 1 pie (9 inches) (22.5 cm)

Crust:
- 1 1/4   **cups (300 ml) graham cracker crumbs**
- 3   **tablespoons (45 ml) sugar**
- 1/3   **cup (75 ml) butter or margarine, melted**
- 1/2   **cup (125 ml) semisweet chocolate morsels**

Filling:
- 1   **package (8 ounces) (226 g) cream cheese**
- 1/4   **cup (50 ml) firmly packed brown sugar**
- 1/2   **teaspoon (2 ml) vanilla extract**
- 1   **cup (250 ml) whipping cream, whipped**
- 1   **pint (2 cups) (500 ml) washed, hulled and sliced fresh strawberries; reserve 1 whole perfect berry for garnish**
- 2   **tablespoons (25 ml) semisweet chocolate morsels**
- 1   **teaspoon (5 ml) butter or margarine**

Preheat oven to 325 degrees F (160 C). Spray 9-inch (22.5 cm) pie plate with nonstick cooking spray.

To make crust, combine cracker crumbs, sugar and butter. Mix well and press firmly onto sides and bottom of prepared pie plate.

Bake 10 minutes. Cool completely.

Melt 1/2 cup (125 ml) chocolate morsels in microwave or double boiler. Stir and cool slightly; drizzle over baked crust.

To make filling, beat cream cheese on medium speed with an electric mixer until light and fluffy. Add brown sugar and vanilla, mixing well. Fold whipped cream into cream cheese mixture; spoon into prepared crust. Refrigerate 8 hours.

Arrange sliced strawberries over filling. Combine 2 tablespoons (25 ml) chocolate morsels and butter. Melt in microwave or in saucepan over low heat; stir and drizzle over strawberry layer. Put reserved berry on top.

Strawberry Butter Pecan Pie

Use fresh strawberries

submitted by Carolyn Beinlich

Farm Wife News *is credited for this strawberry pie with a buttery, nut-flavored, crunchy crust. It's bound to get applause.*

Yield: 1 pie (9 inches) (22.5 cm)

Butter pecan crust:

1/2	**cup (125 ml) butter**
1	**cup (250 ml) flour**
1/4	**cup (50 ml) light brown sugar**
1/3	**cup (75 ml) chopped pecans**

Strawberry filling:

1-11/2	**quarts (1-1.5 L) fresh strawberries, washed and hulled; divided**
3/4	**cup (200 ml) water**
1	**cup (250 ml) sugar**
21/2	**tablespoons (32 ml) cornstarch**

Whipped cream or whipped topping for garnish

Preheat oven to 375 degrees F (190 C). Lightly spray 9-inch (22.5 cm) pie plate with nonstick cooking spray.

To make crust, using fork or pastry blender, cut butter, flour and sugar into coarse crumbs. Stir in pecans. Press mixture into prepared pie plate. Bake 12-15 minutes or until golden brown. Cool.

To make filling, crush 1 cup of the strawberries. Slice remaining strawberries and reserve in refrigerator.

In small saucepan, combine crushed berries, water, sugar and cornstarch. Cook on medium heat until mixture thickens; remove from heat to cool.

Arrange reserved sliced strawberries in crust. Pour cooled glaze over berries. Refrigerate until ready to serve. Top with whipped cream or topping.

We thank clever children for the naming of the sweet red **strawberry**. *After picking the fruit, children strung them on grass straws and sold them "by the straw."*

Home Cooking magazine

Strawberry Desserts

The man in the wilderness
said to me, How many
strawberries grow in the sea?
I answered him as I thought
good, As many red herrings
as grow in the wood.

Mother Goose

Berry Hill Strawberry Shortcake

Use fresh or frozen strawberries

submitted by Doris Stutzman

*Berry Hill Shortcake is credited to Grandma Irwin from Texas
and is served to customers at Stutzman Farms.*

Yield: 8 servings

Shortcake:
- 2 cups (500 ml) flour
- 3 tablespoons (45 ml) baking powder
- 1/2 teaspoon (2 ml) salt
- 3/4 cup (200 ml) sugar
- 1 egg
- 3 tablespoons (45 ml) margarine, melted
- 1 cup (250 ml) milk

Sauce:
- 1 quart (1 L) fresh strawberries washed, hulled and chopped
- 1/2 cup (125 ml) sugar
- Whipped cream, whipped topping and/or ice cream

Preheat oven to 325 degrees F (160 C). Prepare an 8x8-inch (20x20 cm) square pan with nonstick cooking spray.

In a large bowl, sift together flour, baking powder, salt and sugar. Add egg, margarine, and milk; mix well. Bake 25-30 minutes or until light brown. Remove from oven; cool.

To make sauce, in a large bowl, mix together strawberries and sugar. Let stand 30 minutes or until sugar is dissolved.

To serve, cut shortcake into pieces; add ice cream. Spoon strawberry sauce over ice cream. Garnish with whipped cream or topping.

Note: This recipe can be doubled and baked in a 9x13x2-inch (22.5x 30.5x5 cm) pan 40 minutes.

Variation: Substitute frozen sweetened strawberries, thawed for fresh strawberries.

Linnaeus, the Swedish botanist, prescribed for himself a diet of **strawberries**, *and he was cured of the gout—or so he wrote.*

The Strawberry Connection

Deluxe Strawberry Shortcake Sauce

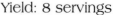

Use fresh strawberries and strawberry jam

submitted by Queensland Fruit and Vegetable Growers

This recipe is adapted from the Queensland, Australia, Fruit and Vegetable Growers newsletter and is truly deluxe.

Yield: 8 servings

1/2	cup (125 ml) strawberry jam
2	teaspoons (10 ml) lemon juice
1	tablespoon (15 ml) brandy (optional)
11/2	quarts (1.5 L) fresh strawberries, washed, hulled and halved; reserve 8 whole berries for garnish
8	Berry Hill Strawberry Shortcakes or shortcakes of your choice
1	cup (250 ml) heavy cream, whipped
	Whipped cream

Combine jam, lemon juice and brandy in a saucepan. Cook and stir over low heat until melted. Cool slightly. Spoon jam mixture over strawberries and refrigerate 1 hour or until set.

Place shortcake on serving plates and cut in half. Spread one half of the whipped cream evenly on shortcake. Spoon strawberry sauce mixture over whipped cream. Garnish with additional whipped cream and whole berry.

 Wife, into thy garden, and set me a plot With **strawberry** *roots, of the best to be got: Such growing abroade, among thorns in the wood Well chosen and picked, prove excellent and good.*

Five Hundred Pointes of Good Husbandrie (1580)

Strawberry Torte

Use frozen strawberries

submitted by Jo Gibbs

This recipe is credited to the Michigan Beekeepers Association.
Serve as a party or holiday dessert; it is well worth the effort.

Yield: 20 servings

Crust:
3/4	cup (200 ml) butter or margarine
1	cup (250 ml) sifted flour
1/4	teaspoon (1 ml) salt
1/2	cup (125 ml) sugar

Meringue topping:
4	egg whites
1/8	teaspoon (.5 ml) cream of tartar
1/8	teaspoon (.5 ml) salt
1/2	teaspoon (2 ml) almond extract
8	tablespoons (120 ml) sugar

Filling:
3	tablespoons (45 ml) cornstarch
3/4	cup (200 ml) sugar
2	packages (10 ounces each) (283 g each) frozen unsweetened strawberries, sliced or chopped and thawed, with juice
1/4	cup (50 ml) honey
1-2	drops red food coloring (optional)

To make crust, preheat oven to 350 degrees F (180 C). In a medium bowl, combine flour, salt and sugar; cut in butter with fork or pastry blender. Press mixture into bottom of 9x13-inch (22.5x30.5 cm) pan. Bake 10-15 minutes or until lightly browned. Cool.

To make filling, mix cornstarch and sugar in saucepan. Stir in strawberries and honey. Bring to a boil over high heat, stirring constantly. Reduce heat to medium and continue cooking until mixture is clear, stirring constantly. Remove from heat and cool slightly. Pour over baked crust.

To make meringue, combine egg whites, cream of tartar, salt and extract; beat until foamy. Add sugar several tablespoons at a time, beating after each addition until meringue stands in peaks. Spread over filling, sealing edges. Return pan to oven and bake 15 minutes or until meringue is light brown.

Variation: If frozen sweetened strawberries are used, reduce sugar to 1/4 cup (50 ml).

Don and Jo Gibbs have been growing strawberries since 1970 on their farm located halfway between Lansing and Jackson, Michigan. Twenty-five acres are currently sold 25 percent pick-your-own and 75 percent retail on the farm and at farmer's markets.

Strawberry Fruit Compote

Use fresh or frozen strawberries

submitted by Dr. Gene Galletta

*Serve this perfect dessert following a heavy pasta, meat or seafood dinner.
Coffee and white wine, such as Asti Spumante or Piesporter,
provide a nice accompaniment to this dessert.*

Yield: 6 servings

1   **quart (1 L) fresh strawberries, washed, hulled and sliced
    or 2 pints (1 L) frozen unsweetened sliced strawberries**
1   **large can (29 ounces) (822 g) cling peach slices, chilled**
2   **teaspoons (10 ml) honey, divided**
2   **teaspoons (10 ml) lemon juice, divided**
2   **bananas, sliced**
Mint sprigs (optional)

Partially thaw strawberries. Drain peaches and reserve juice. In a large glass bowl, add strawberries, peaches, 1 teaspoon (5 ml) each of the honey and lemon juice. Toss lightly and refrigerate until just before serving.

When ready to serve, toss fruit again lightly. Garnish top with banana slices and mint sprigs. Mix and add remaining 1 teaspoon each honey and lemon juice; add peach juice to taste.

Compote may be served in sauce dishes or long stemmed fruit or champagne glasses.

Gene and Nada Galletta live and garden in suburban Laurel, Maryland. Gene breeds berries, especially strawberries, at the nearby Beltsville Agricultural Research Center.

Aunt Jean's Strawberry Betty

Use fresh or frozen strawberries

submitted by Phyllis Schartner

This old-fashioned family recipe is sure to please a modern-day cuisine.

Yield: 8 servings

2	tablespoons (25 ml) butter, melted
1 1/2	cups (375 ml) soft bread crumbs, divided
2	cups (500 ml) washed, hulled and sliced fresh strawberries
1/2-1	cup (125-250 ml) sugar (depending on ripeness and acidity of strawberries)
1	lemon; juice squeezed and reserved, peel grated
1/4	teaspoon (1 ml) cinnamon (optional)
1/4	teaspoon (1 ml) nutmeg (optional)
1/4	cup (50 ml) water

Preheat oven to 350 degrees F (180 C). Prepare an 8x8-inch (20x20 cm) with nonstick cooking spray.

In a small bowl, pour butter over bread crumbs; stir to mix. Layer 3/4 cup of the bread crumbs, strawberries, sugar, lemon juice, grated lemon peel, cinnamon and nutmeg into prepared baking dish. Sprinkle top with remaining 3/4 cup bread crumbs. Pour water over top.

Cover pan with aluminum foil to prevent excessive overbrowning and bake 20 minutes; remove foil and continue baking for approximately 25-30 more minutes.

Variation: Substitute 2 cups (500 ml) frozen unsweetened sliced strawberries or 2 packages (10 ounces each) (284 g each) frozen sweetened strawberries, drained. Reserve juice and pour over top in place of water; omit sugar.

 Herb and Phyllis Schartner grow strawberries (retail and wholesale), apples, raspberries and plums in scenic, rural north central Maine—that part of Maine some people only read about. They advertise "experience an event." This includes the fruit express and horse-drawn hay rides, which bring people from all over for the best picking available.

Agutaq (Eskimo Ice Cream)

Use fresh strawberries

submitted by Agnes Bostrom

*Agutaq is a favorite food of the Alaskan natives. When available,
traditional seal oil, reindeer tallow and wild berries from the tundra are still used.*

Yield: 8 servings

2	pounds (960 g) fresh sheefish, whitefish or northern pike
1 1/2	cups (375 ml) Crisco shortening
1	cup (250 ml) sugar
1	tablespoon (15 ml) vegetable oil
1/2	cup (125 ml) raisins
2	quarts (2 L) fresh strawberries, washed, hulled and chopped (not crushed)

Cut fish in 3-inch (7.5 cm) pieces. In a saucepan, cover fish with water and bring to a boil over high heat; reduce heat to medium and continue to boil gently for 20 minutes.

Remove from heat; drain and cool until fish can be handled. Remove skin and bones; squeeze and remove fluids. Flake fish to equal 3 cups. Rub fish between hands to a very fine consistency.

With an electric mixer or food processor, cream together shortening, sugar, and oil until light and fluffy. Gradually add fish, maintaining fluffy consistency. If mixture is too dry, add 1 tablespoon (15 ml) water at a time. Fold in raisins and strawberries until well mixed. Refrigerate.

Serve fresh or keep frozen until ready to serve. Garnish with a whole fresh strawberry.

Ed and Aggie Bostrom of North Pole Acres in Alaska, grow **Toklat** *strawberries, rhubarb, asparagus, annual vegetables, flowers and hay. They produce compost, potting soil, container plants, 32 varieties of jams; jellies, syrups, butters and dessert toppings to sell at markets and bazaars. Native Eskimo arts, crafts and apparel are also available. Bed and breakfasts, bus tours and pick-your-own strawberries complete their operation.*

Kid's Frozen Pudding Treats

Use fresh or frozen strawberries

submitted by Marilyn Gjevre

Yield: 15 pudding treats

1   **package (3.4 ounces) (96 g) instant vanilla pudding**
1   **cup (250 ml) milk**
1   **container (8 ounces) (226 g) whipped topping, thawed
    (LaCreme with real creme recommended)**
1   **pint (500 ml) fresh strawberries, washed, hulled and mashed**
1   **cup (250 ml) mashed bananas**
15 **paper drinking cups (5 ounces each) (150 g each)**
15 **wooden popsicle sticks**

With hand beater or electric mixer, prepare vanilla pudding according to directions using 1 cup (250 ml) milk. Fold in whipped topping, mashed strawberries and banana. Spoon into paper drinking cups and insert wooden sticks. Freeze 4 hours or until firm. To serve, remove from freezer and peel off paper cups.

Note: Popsicle molds or ice cube trays can be substituted for paper cups.

Variation: Substitute chocolate or desired flavor of instant pudding for vanilla pudding; use 1 package (16 ounces) (453 g) frozen unsweetened whole strawberries, partially thawed and mashed for fresh strawberries.

Frozen Strawberry-Yogurt Sandwiches

Use fresh strawberries

submitted by Jo Gibbs

Yield: 9 sandwiches

1   **quart (1 L) frozen vanilla yogurt**
3   **cups (750 ml) washed, hulled and sliced fresh strawberries**
18 **graham cracker squares, divided**

Soften frozen yogurt by placing in a large bowl; break up and stir with spoon. (Do not allow to melt.) Fold in strawberries.

Arrange 9 of the graham crackers in an 8-inch (20 cm) square pan. Spread yogurt-strawberry mixture over crackers. Cover with remaining 9 crackers to make sandwiches. Freeze until firm. Cut into 9 squares.

Variation: Substitute ice cream or frozen non-fat yogurt for regular yogurt.

Strawberries Elegante

Use fresh strawberries

submitted by Eldon Galloway

Besides being elegant, this dessert is light and easy!

Yield: 8 desserts

6   cups (1500 ml) washed, hulled and sliced fresh strawberries; reserve 8 whole berries with stem
3   tablespoons (25 ml) orange juice or orange liqueur, divided
1   package (8 ounces) (227 g) light cream cheese
3   tablespoons (45 ml) brown sugar
1   tablespoon (15 ml) skim milk

In a medium bowl, marinate sliced strawberries with 2 tablespoons of the orange juice for 1 hour.

Combine cream cheese, brown sugar, remaining 1 tablespoon orange juice and milk in blender; blend until smooth.

To serve, put berries in individual dishes or glass goblets and spoon or layer with cream cheese mixture. Garnish with whole strawberry with stem or a mint leaf.

Berried Treasures is a family-owned business in Fort Saskatechewan, Alberta, Canada, that has been growing strawberries since 1989. The Galloways thrive on giving their customers a very personal experience that keeps them coming back for more . . . "great service and the 'best tasting strawberries in the world.'"

Strawberry-Wine Ice

Use fresh or frozen strawberries

submitted by Bea Statz

*This recipe can be made during the strawberry season
and will keep in the freezer for months. It makes a refreshing,
light dessert after a heavy meal—especially during the holidays!*

Yield: 12 servings

1	**quart (1 L) fresh strawberries, washed and hulled or frozen unsweetened strawberries**
2	**cups (500 ml) Rhine wine, divided**
3/4	**cup (200 ml) honey**
1	**teaspoon (5 ml) lemon juice**

Combine strawberries, 1 cup (250 ml) of the wine, honey and lemon juice in blender or food processor. Blend on high until very smooth. Add remaining 1 cup wine; blend until well mixed. Pour into 3 ice cube trays. Freeze until solid.

Right before serving, place 1 tray of frozen ice cubes into blender. Process until mixture becomes a fine ice with a slightly granular texture. If necessary, stop machine and push cubes to the bottom so that all are mixed evenly. Continue with remaining 2 trays as needed. Spoon into glass sherbet dishes and serve at once. Garnish with fresh whole strawberry with hull.

Note: During strawberry season this recipe can be made and frozen in large containers. To serve, scrape frozen mixture with a spoon to create a fine ice. Remove as much as needed by scraping so remaining ice stays frozen solid. It will keep for up to 1 year.

*The **strawberry** was one of the crops grown by Iroquois Indians and important enough to be center stage in many festivals held in its honor.*

Lemon Strawberry Sorbet

Use fresh or frozen strawberries

submitted by Mary Secor

*Adapted from an employee's recipe calling for fresh peaches
and orange juice, this lemonade and strawberry version will provide
a refreshing, cool dessert all year-round.*

Yield: 1 quart (1 L)

   1   **cup (250 ml) buttermilk**
 1/4 **cup (50 ml) frozen lemonade concentrate**
   3   **cups (750 ml) fresh strawberries, washed and hulled
        or frozen unsweetened whole strawberries**
   1   **cup (250 ml) sugar**

In a blender, combine buttermilk, concentrate, strawberries and sugar. Blend until smooth. Pour into covered plastic bowls or ice cube trays and freeze.

To serve, scrape with spoon from bowl of frozen sorbet or put cubes into blender to mix until smooth. Garnish with fresh whole strawberries or mint leaves.

 The French called the **strawberry** *"fresas" because of the excellent sweetness and odor.*

Anonymous (1536)

Strawberries Savannah

Use fresh strawberries

submitted by Carolyn Beinlich

A cool, easy summer dessert recipe adapted from **Sphere** *magazine.*

Yield: 12 servings

3	**pints (1.5 L) fresh strawberries, washed, hulled and cut in half**
1/3-1/2	**cup (75-125 ml) powdered sugar**
1/3	**cup (75 ml) orange juice or orange liqueur**
1/2	**cup (125 ml) whipping cream**
1/2	**cup (125 ml) dairy sour cream**
2	**tablespoons (25 ml) powdered sugar**
1/2	**teaspoon (2 ml) ground mace**

In a bowl, mix strawberries, powdered sugar and orange juice. Refrigerate 2 hours.

Measure whipping cream, sour cream, powdered sugar and mace into chilled mixing bowl. Beat with electric mixer until fluffy. Refrigerate no longer than 30 minutes.

To serve, spoon strawberries into individual sherbet dishes and top with cream mixture.

 *There are certain foods, like the **strawberry**, that carry a beautiful punch because they're loaded with the vitamins and minerals that every woman needs to look her very best. The **strawberry** is vital for super skin. Count on the pick of the berry crop for Vitamin C, which helps form collagen, the supportive tissue needed to keep skin firm and smooth. In fact, an average-size one-cup serving of **strawberries** supplies more Vitamin C than an average-size orange (88 versus 66 milligrams). **Strawberries** also dish up Vitamin A, which is essential for healthy hair and skin and contain potassium. At 55 calories a cupful, they're great if you're watching your weight.*

"Fitness and Nutrition," *The Minneapolis Times*

Strawberry-Rhubarb Crunch

Use fresh or frozen strawberries

submitted by Doris Stutzman

*Make when fresh strawberries and rhubarb are in season
or use frozen for this baked dessert. They have great complementary flavors.*

Yield: 8 servings

Crust:
- 1 **cup (250 ml) flour**
- 3/4 **cup (200 ml) oatmeal**
- 1 **cup (250 ml) brown sugar**
- 1/2 **cup (125 ml) butter, melted**
- 1 **teaspoon (5 ml) cinnamon**
- 2 1/2 **cups (625 ml) fresh or frozen diced rhubarb**
- 1 1/2 **cups (375 ml) fresh or frozen strawberries**

Filling:
- 1 **cup (250 ml) sugar**
- 1 **cup (250 ml) water**
- 2 **tablespoons (25 ml) cornstarch**
- 1 **teaspoon (5 ml) vanilla extract**

Preheat oven to 350 degrees F (180 C). Prepare a 9x9-inch (22.5x22.5 cm) square pan with nonstick cooking spray.

To make crust, in a large bowl, add flour, oatmeal, brown sugar, butter and cinnamon. Using a fork or pastry blender, mix until crumbly. Press half of the mixture into prepared pan. Layer rhubarb and strawberries over crust; set aside.

To make filling, in a saucepan combine sugar, water and cornstarch. Cook and stir over medium heat until thick and clear. Add vanilla; stir until well blended. Pour over fruit mixture. Top with remaining half of crumbs. Bake 50-60 minutes. Remove from oven.

Cut into squares when cool, adding a scoop of ice cream, if desired.

 This berry is the wonder of all the fruits. One of the chiefest doctors of England was wont to say that God could have made, but God never did make, a better berry. The Indians bruise them and make **strawberry** *bread.*

Roger Williams (1643)

Strawberry Dessert Roll

Use fresh strawberries

submitted by Virginia Radewald

This light, tasty and pretty party dessert is good for any luncheon or dinner.

Yield: 8 servings

5	**eggs, separated**
1	**cup (250 ml) sugar, divided**
1	**cup (250 ml) flour**
1	**teaspoon (5 ml) baking powder**
1/2	**teaspoon (2 ml) salt**
1	**teaspoon (5 ml) vanilla extract**
1	**quart (1 L) fresh strawberries, washed, drained, hulled and sliced; reserve whole strawberries for garnish**
2	**cups (500 ml) whipped cream or whipped topping**

Preheat oven to 375 degrees F (190 C). Line a 10x15-inch (25x37.5 cm) jelly roll pan with waxed paper.

Beat yolks with 1/2 cup (125 ml) of the sugar until thick and lemon-colored. Add vanilla and beat well.

In a separate bowl, beat egg whites until almost stiff. Add remaining 1/2 cup (125 ml) sugar and beat until very stiff. Fold egg yolk mixture into egg white mixture.

Sift together flour, baking powder and salt. Fold into egg mixture. Spread batter into prepared pan; bake 12 minutes.

Remove from oven and immediately turn cake onto a linen towel that has been sprinkled with powdered sugar. Peel off waxed paper and place a new piece on cake. Roll up as a jelly roll and cool 30 minutes or more.

Unroll and remove waxed paper. Spread whipped cream on cake. Spoon sliced strawberries over whipped cream. Roll up and refrigerate until ready to serve. Cut in slices and garnish with whipped cream and whole strawberries. (Whole roll may also be frosted with whipped cream before slicing.)

Variation: Prepare an angel food cake mix as directed on package and bake in jelly roll pan at 350 degrees F (180 C) 15-20 minutes. Continue to follow recipe instructions.

 *A native Hawaiian **strawberry** was discovered over 150 years ago by a French horticulturist. The discovery was accidental. The fruit was called "Ohelo," meaning a pink-colored berry.*

The Strawberry Connection

Strawberry Jams, Jellies and More

The thing about farming is
you are responsible for a
small portion of the planet.
You keep it or you lose it and
it changes, no matter what,
according to your efforts. But
what's important is when you
wake and pick **strawberries**,
not only are they good—
they're yours.

Michael Carey,
"The Thing About Farming,"
The Noise the Earth Makes

Strawberry Jams, Jellies and More

This chapter contains suggestions for making jams and jellies with sugar. See the "Sugarless, Low-Fat and Low-Calorie Choices" chapter for jam recipes without sugar. It also includes recipes with or without commercial fruit pectin. There are recipes for Strawberry Syrup (page 94) and Strawberry Leather (page 95) plus suggested ways to dry strawberries—all alternate methods for preserving strawberries.

Using Pectin

Since strawberries naturally have only a small amount of pectin, using commercial fruit pectin is recommended. The yield will be greater and the product will be more reliable. The disadvantage seems to be in the taste, as many people feel the jam tastes more of sugar than of fruit. Each brand and type of commercial fruit pectin has a different formulation. Making jams and jellies is an exact science and precise amounts of sugar, fruit and acid are necessary for a good set. **Purchase the brand of pectin of your choice and make no substitutions, following package instructions carefully.**

The major brands of fruit pectin are: SURE•JELL Fruit Pectin, SURE•JELL for Lower Sugar Recipes Fruit Pectin and CERTO Liquid Fruit Pectin made by Kraft Foods, Inc., and BALL 100% Natural® Fruit Jell™ Regular Pectin and BALL 100% Natural Fruit® Jell™ No Sugar Needed Pectin. Kraft Foods and Ball include kitchen-tested strawberry recipes with their products. Be sure to read and follow all instructions included in the package. There may also be other good commercial fruit pectin on the market in your area. **The important thing is to carefully follow the recipe included in the package.**

Jams and Jellies with Pectin

The most popular strawberry jam is Basic Strawberry Freezer Jam. Follow the recipe in the fruit pectin package or try one of these flavorful variations:

Basic Strawberry Freezer Jam Variations:

Any one of the following may be added to berries before adding sugar: 1/2 cup (125 ml) slivered almonds, chopped pecans, pumpkin seeds, sunflower seeds or pine nuts (toasted, if desired); 1/4 cup (50 ml) dry white vermouth or sherry wine; 2 tablespoons (25 ml) orange liqueur; 1 tablespoon (15 ml) finely chopped crystallized ginger; 2 teaspoons (10 ml) almond extract; 1 tablespoon (15 ml) grated lemon, lime or orange rind; 1/4 cup (50 ml) mashed ripe banana and 1 tablespoon (15 ml) lemon juice.

Strawberry Jelly

Use fresh strawberries

Adapted from a recipe and used with permission from
BALL 100% Natural® Fruit Jell™ Regular Pectin

This recipe produces a clear red jelly with no seeds.

Yield: 8 cups (2 L)

3   quarts (3 L) fully ripe strawberries
4 1/2   cups (1.125 L) sugar
1   box BALL 100% Natural® Fruit Jell™ Regular Pectin

Wash, drain, hull and crush strawberries one layer at a time in a large bowl. Place crushed berries in dampened jelly bag and let juice drip for several hours over 6- to 8-quart (6-8 L) saucepan. For clear juice do not squeeze bag.

Measure 3 1/2 cups (875 ml) prepared juice. If yield is short, add water to fruit pulp and extract additional juice.

Measure sugar into medium bowl; set aside.

Stir fruit jell into prepared juice. Bring to a full boil over high heat, stirring constantly. Add sugar; return to a full, rolling boil. Boil hard 1 minute, stirring constantly. Remove from heat. Skim off foam, if necessary. Quickly ladle into 8, 1-cup (250 ml) hot sterilized jars, leaving 1/4-inch (1 cm) headspace. Adjust caps.

Process all jars 5 minutes in boiling water-bath. Remove jars from canner. Let jelly cool 12-24 hours. Check lids for seal. Remove bands and clean exterior surface of jars and lids. Store jelly in a cool, dry, dark place for up to 1 year.

Note: Strawberry Jelly recipes can also be found with other brands and types of pectin. Follow recipe included with package directions.

Jams and Jelly Troubleshooting

Ellen Todd, NASGA member, compiled an informational handout which lists answers to the "most frequently asked" questions:

Why is my strawberry jam or jelly too soft?

A common problem is jam or jelly has not set, resulting in a soft or runny product. This problem can be caused by a number of factors, including:

1. too much juice in the mixture
2. not enough sugar
3. mixture not acidic enough
4. made too large a batch

Sometimes a too-soft product can be improved by recooking 4 to 6 cups (1-1.5 L) jam or jelly at a time. Use the following guidelines to correct your jam or jelly, or use unset product as a syrup over ice cream, pancakes, custard or cheesecake.

To reset with powdered fruit pectin: Measure jam or jelly to be recooked. For each quart (1 L), measure 1/4 cup (50 ml) sugar, 1/4 cup (50 ml) water and 4 teaspoons (20 ml) powdered pectin. Mix pectin and water; bring to a boil, stirring constantly to prevent scorching. Add jam or jelly and sugar. Stir thoroughly. Bring to a full, rolling boil over high heat, stirring constantly. Boil hard for 30 seconds. Remove from heat; skim and pour into sterilized jars. Seal with new lids. Process in hot water-bath canner 5 minutes or invert jars 5 minutes.

To reset with liquid pectin: Measure jam or jelly to be recooked. For each quart (1 L), measure 3/4 cup (200 ml) sugar, 2 tablespoons (25 ml) lemon juice and 2 tablespoons (25 ml) liquid pectin. Bring jam or jelly to a boil over high heat, stirring constantly to prevent scorching. Quickly add sugar, lemon juice and liquid pectin; bring to a full, rolling boil. Stir constantly and boil mixture 1 minute. Remove from heat; skim and pour into sterilized jars. Process in hot water-bath canner 5 minutes or invert jars 5 minutes.

Strawberry Jam Using Frozen Strawberries

It is possible to make either freezer or cooked jam from frozen unsweetened strawberries. This is recommended for those who do not have the time to make jam during the local strawberry season or for those who run out of homemade jam before next season begins. Making jam from frozen berries is a great way to use last season's berries from the freezer.

The secret to making jam from frozen strawberries is to allow the strawberries to thaw and be brought to room temperature before beginning to make jam or jelly. Crush berries one layer at a time. Do not leave whole pieces of fruit. Read and carefully follow directions included with the fruit pectin being used.

Jams and Jellies Without Pectin

Making cooked jams and jellies without pectin results in a richer fruit flavor. The disadvantage of making jams, jellies and preserves without pectin is that it is a less reliable method and the yield will be less. When making cooked preserves without pectin, the secret is to boil the mixture for a longer period of time until the jelling stage is reached. Candy thermometers are necessary; cook until it registers between 220-225 degrees F (104-106 C).

If a candy thermometer is not available, another way to test for the jelling stage, although it is less reliable, is to use the metal spoon test. Dip a metal spoon into the cooking preserves, lift it out, and watch the liquid fall from the spoon. When it gets syrupy, the liquid will fall away into two drops, side by side. When the two drops begin to come together and form a sheet as they fall from the spoon, the jelling stage has been reached.

Old-Fashioned Whole Berry Preserves

Use fresh strawberries

submitted by Doris Stutzman

This is an old-fashioned recipe like grandma made but with more accuracy.

Yield: 6 cups (1.5 L)

6 **cups (1.5 L) fresh small fully ripe whole strawberries,
    washed and hulled**
Boiling water to cover strawberries
1/2 **cup (125 ml) lemon juice**
6 **cups (1.5 ml) sugar, divided**

In a large saucepan, cover berries with boiling water; let stand 3 minutes to soften. Drain water and discard.

Combine berries and 3 cups (750 ml) of the sugar in a 6- to 8-quart saucepan. Bring to a boil over high heat, stirring constantly. Reduce heat; continue to boil slowly on medium heat 8 minutes, stirring constantly. Add remaining 3 cups (750 ml) sugar and lemon juice. Boil 10 minutes more, stirring constantly. Using a candy thermometer, bring to the jelling stage of 220-225 degrees F (104-106 C).

Remove from heat. Stir and skim off foam with metal spoon for 2 minutes. Pour jam into shallow baking dish to cool completely. When cold, put in sterilized jars and freeze or process 10 minutes in boiling water-bath to seal lids.

The fruit sugar of **strawberries** *is more easily assimilated by people suffering from mild diabetes than is any other type of sugar. The salts are laxative, and with the seeds and fiber will stimulate sluggish bowels. The mineral salts present in the* **strawberry** *contain a high proportion of potassium, phosphoric acid and iron, so there is some scientific basis for the tradition that* **strawberries** *have some tonic effect. The iron helps make the fruit red.* **Strawberries** *are remarkably soluble in solutions similar to those of the alkaline digestive juices of the intestines: it can then be assumed with some confidence that they are good for digestion.*

The Compleat Strawberry

Strawberry Syrup

Use fresh strawberries

submitted by Bea Statz

This recipe appeared in **Parade** *magazine. This scarlet syrup is a must for ice cream sodas, pancakes and French toast.*

Yield: 2 cups (500 ml)

4   cups (1 L) fully ripe strawberries, washed and hulled
1 3/4   cups (450 ml) water, divided
2   teaspoons (10 ml) finely grated lemon peel
1 1/4   cups (300 ml) sugar

Crush berries in a heavy saucepan. Add 1 cup (250 ml) of the water and lemon peel. Bring to a boil. Reduce heat slightly and simmer over medium heat 5 minutes. Skim foam from top with metal spoon. Set mixture aside to cool.

Meanwhile, combine sugar and remaining 3/4 cup (200 ml) water in a small heavy saucepan. Bring to a boil and cook until the syrup reaches 260 degrees F (110 C) on a candy thermometer. Set mixture aside.

Strain cooled strawberry mixture through a double thickness of cheesecloth. Squeeze until all juice and pulp are extracted and seeds are left behind. Discard cheesecloth contents.

Pour clear, strained liquid into the heavy saucepan with sugar-syrup mixture. Bring to a boil and cook 8 minutes, stirring constantly. Pour into 2, 8-ounce (250 ml) jars, leaving 1/4 inch (1 cm) of headspace. Process in boiling water-bath 10 minutes or invert jars for 5 minutes.

 In 1848 the Massachusetts Horticultural Society held its first straw-berry (bold) exhibition. The first **strawberry** *festival in America was held in the Boston suburb of Belmont, Massachusetts, a decade later. After that, the custom of giving* **strawberry** *parties and holding* **strawberry** *festivals took the country by storm and continues today.*

Old-Fashioned Strawberry Recipes With Historic Notes

Strawberry Leather

Use fresh strawberries

submitted by Doris Stutzman

Strawberry Leather is a good snacking treat. The kids will love it!

Yield: 2 jelly roll pans

5 cups (1.25 L) washed, hulled and halved fresh strawberries
1/4 cup (50 ml) sugar or honey

Preheat oven to 150 degrees F (65 C). Line 2 jelly roll pans with plastic wrap. Secure edges with tape. Place berries one cup at a time in blender. Puree until smooth. Blend in sugar. Spread mixture evenly in pans.

Place in oven leaving door ajar approximately 4 inches (10 cm). Place oven thermometer in back of oven and check periodically that temperature registers 150 degrees. If necessary, turn oven off for awhile to keep at 150-degree temperature. Rotate pans every 2 hours. The leather is dry when surface is no longer sticky. Slice into strips and store in an air-tight container in a cool, dry place.

Drying Strawberries

Use fresh strawberries

submitted by Doris Stutzman

Yield: 2 cups (500 ml)

Fresh, small strawberries, washed, hulled and drained on paper towel (can be halved if large berries)

Method No. 1: If a food dehydrator is available, process according to directions included with dehydrator.

Method No. 2: A homemade dryer rack made from clean screening can be used in a dry, airy place. Dry for several days or until strawberries are firm and no juice is evident.

Method No. 3: For oven-dried strawberries, preheat oven to 150 degrees F (65 C). Place strawberries in oven on greased cookie sheet. Check often to be sure oven is not too hot. Leave oven door open if necessary. Dry 2-4 hours or until strawberries are firm and no juice is evident.

To store dried strawberries, place in jar and cover until ready to use.

The Strawberry

by H. H. Harris
(reprinted from 1930 Wisconsin Strawberry Field Day Program.)

Some will lack the patience
Say that strawberries never grow
That will pay for all the labor
That on them you must bestow.

But come and listen while I tell you
Of the pleasure that I find
In a patch of well-kept strawberries
Watched and cared for to my mind.

And methinks I read a promise
In the Merry Month of May
Your fond hopes shall find fruition
Every blossom seems to say.

See the plants shine in the sunlight
Plants of kinds both old and new
promising to shade my berries
With their leaves of richest hue.

Days pass by 'mid cloud and sunshine
Hum of bee and song of bird
And, ere long, from dawn till sunset
Pickers cheerful tones are heard.

And the well-filled crates and baskets
Each one packed with skill and care
Daily ar borne off to market
Through all weather, foul and fair.

But 'mid all the summer's pleasure
For the winter we provide
And a glance within our pantry
Shows you standin side by side

Jars of sweetmeats, jams and jellies
From the luscious berries made
Gathered by the busy household
Ere the days of ripening stayed.

When with the toil we've grown weary
To the festive board we come
Here we find the toothsome shortcake
In our own beloved home.

Spread with ripe and juicy berries
Served with richest, sweetest cream
And a dish of fine, ripe strawberries
All will vanish like a dream.

And we know our Heavenly Father
Smiles upon the path we've trod
We within the book of nature
See the hand of Nature's God.

NASGA Strawberry Farms—1998

For membership information contact:
Robert and Donna Cobbledick • NASGA Executive Secretaries
324 Lake Street • Grimsby, Ontario, Canada L3M 1Z4
Phone: (905) 945-9057 • FAX: (905) 945-8643

Legend: ***recipe contributor**.
Contact the strawberry farm in your area for more information.

UNITED STATES
(in alphabetical order by state)

ALASKA
Bostrom, Agnes (Aggie)*
Ed and Agnes Bostrom
North Pole Acres
North Pole, AK 99705

ARKANSAS
Cross, Bill and Susan
Harvest Fresh Farm
McNeil, AR 71752

CONNECTICUT
Lyman, John III
The Lyman Farm Inc.
Middlefield, CT 06455

Brown, Allyn III
Maple Lane Farms
Preston, CT 06365

Jones, Terry and Jean
Jones Tree Farm
Shelton, CT 06484

Dzen, Donald Jr. & Joseph
Dzen Farms Inc.
South Windsor, CT 06074

Orr, Peter and Kristin
Fort Hill Farms
Thompson, CT 06277

Pell, Roger & Luping
Pell Farms
Somers, CT 06071

Donnelly, Tom and Carol
Pleasant Valley Farm
Dayville, CT 06241

Martin, Kathi
Brown's Harvest
Windsor, CT 06095

FLORIDA
Brown, Marvin and Linda
Favorite Farms Inc.
Dover, FL 33527

GEORGIA
Calhoun, Gerald & Joyce
Calhoun Produce Inc.
Ashburn, GA 31714

Mathews, Bill and Amogene
Mathews Farms
Baxley, GA 31513

IDAHO
Haroldsen, Grant and Sue
St. Leon Pick-Your-Own
Produce
Idaho Falls, ID 83401

ILLINOIS
Halat, Cheryl*
Tom and Cheryl Halat
Tom's Vegetable Market
Huntley, IL 60142

Stanley-Ramos, Linda*
Produce Promotions
Chicago, IL 60660

INDIANA
Johnson, Rodney and Jenny
Johnson's Farm Produce
Hobart, IN 46342

Erwin, Sam
Pickin' Patch, Inc.
Plymouth, IN 46563

Wright, Warren
Wright's Berry Farm
Newburgh, IN 47630

Joe Huber
Joe Huber Family
Farm\Orch\Rest.
Borden, IN 47106

Ahrens, Phil and Linda
Ahrens' Garden of Eatin' Inc.
Huntingsburg, IN 47542

Roseberry, Marlin and Elyse
Rosey's Berry Farm
Michigan City, IN 46360

IOWA
Maahs, Gene and Naomi
Country Gardens
Adel, IA 50003

Furleigh, Robert and Donna
Furleigh Farms
Clear Lake, IA 50428

KENTUCKY
Hieneman, Curtis
Hieneman Farm
Greenup, KY 41144-9709

Courter, Treva*
Dr. J. W. and Treva Courter
Kevil, KY 42053

Blake, Cal and Judi
Strictly Strawberries
Lexington, KY 40515

MAINE
Fenimore, Donald and Pat
Fenimore's Strawberry Farm
Bowdoin, ME 04008

Bucknell, Roger and Franklin
Bucknell Farm
Denmark, ME 04022

Maxwell, Elsie L.*
Kenneth and Elsie L. Maxwell
Maxwell's Farm
Cape Elizabeth, ME 04107

Gleason, Robert
Green Point Farms
Dresden, ME 04342

Tate, Ken and Bev
Tate's Strawberry Farm
East Corinth, ME 04427

Adams, Joseph and Carol
Adams Strawberry Acres
East Corinth, ME 04427

Chipman, Ellsworth & Douglas
Chipman Farms
Poland Springs, ME 04274

Stevenson, Ford and Susan
Stevenson's Strawberries
Wayne, ME 04284

Pike, David and Verna
Farmington, ME 04938

Bernier, Donald and Pauline
Bernier Egg Farms Inc.
Sanford, ME 04073

Popp, David
Popp Farm
Dresden, ME 04342

Schartner, Phyllis*
Herb and Phyllis Schartner
Mt. View Fruit & Berry Farm
Thorndike, ME 04986-0082

MARYLAND
Galletta, Dr. Gene*
Fruit Laboratory USDA-ARS
Beltsville, MD 20705

Johnson, Philip and Ruth Ann
Walnut Springs Farm
Elkton, MD 21921

Butler, Susan*
Susan, George & Shirley Butler
Butler's Orchard
Germantown, MD 20876

Biggs, Richard and Nancy
Rock Hill Orchard
Mt. Airy, MD 21771

Allnutt, Benoni and Maureen
Homestead Farm
Poolesville, MD 20837

Moore, Guy and Lynn
Larriland Farm
Woodbine, MD 21797

MASSACHUSETTS
Parlee, Ralph
Parlee Farms
Chelmsford, MA 01824

Valonen, Rudy
Valonen's Berry Farm
East Longmeadow, MA 01028

Oathout, James Jeffery
Marlborough, MA 01752

Tougas, Maurice & Phyllis
Tougas Family Farm
Northborough, MA 01532-1006

Moskow, Bencion and Patricia
Thimble Farm
Vineyard Haven, MA 02568

Rogers, Richard and Betty
Roger's Spring Hill Farm, Inc.
Ward Hill, MA 01835

Marini, V. Mario
Marini Farm
Ipswich, MA 01938

Parlee, Mark and Ellen
Parlee Farm
Tyngsboro, MA 01879

MICHIGAN
Tervo, Thomas J.
Tervo's Aching Acres
Chassell, MI 49916

Long, Rob and Chris
Long Family Orchard & Farm
Commerce, MI 48382

Pellergrini, Don and Patricia
Pellegrini's Strawberry Farm
Escanaba, MI 49829

Symanzik, Eugene and Coralyn
Symanzik's Berry Farm
Goodrich, MI 48438

DeGroot, Roy and Marsha
DeGroot's Strawberries
Gregory, MI 48137

DeLange, Phil and Theresa
Redberry Farm
Hudsonville, MI 49426

Bardenhagen, Gary & Christi
Lake Leelanau, MI 49653

Altermatt, Jack and Darlene
Altermatt's U-Pick Farm
Macomb, MI 48042

Deneweth, Ken and Carolyn
Deneweth Greenhouse & U-Pick
Macomb, MI 48044-0128

Robinson, Thomas & Jo Ann
S & S Farm Market
Middleville, MI 49333

Kudwa, Chester
Idlewild Farms
Crystal Falls, MI 49920

Ramey, Roy and Maria
Ramey Farms
New Era, MI 49446

Radewald, Virginia*
Edwin and Virginia Radewald
Radewald Farms
Niles, MI 49120

Middleton, William and Barb
Middleton Berry Farm
Oakland, MI 48363

Gibbs, Jo*
Donald and Jo Gibbs
Don Gibbs Farm
Onondaga, MI 49264

Blake, Ray and Jan
Blakes Orchard and Cider Mill
Armada, MI 48005

Mieske, Melvin and Deloris
Monarch Farms
Midland, MI 48652

Bigelow, Marvin
Bigelow Berry Farm Inc.
North Branch, MI 48461

MINNESOTA
Ware, Bernard Jr.
Ware Farm
Bear Lake, MN 49614

Kringen, Darrell and Georgia
Kringen's Strawberry Patch
Mora, MN 55051

Peterson, Donald L.
Peterson's Strawberries
Rosemount, MN 55068

Edberg, Kevin
The Berry Patch
White Bear Lake, MN 55110

Jacobson, Bill
Pine Tree Apple Orchard
White Bear Lake, MN 55110

Carman, Richard and Margaret
Carman Berry Farm
Wadena, MN 56482

Lodermeier, Duane and Diane
Lodermeier Gardens
St. Cloud, MN 56303

Hassett, Bob and Nancy
Hassett's Berry Farm
Big Lake, MN 55309

MISSOURI
Stephenson, Ronald
Stephenson's Orchards
Kansas City, MO 64136

MISSISSIPPI
Johnson, James E.
Tupelo, MS 38801

Maust, Elaine*
Duane and Elaine Maust
Sunshine Berry Farm
Meridian, MS 39307

NEW HAMPSHIRE
Patenaude, Wayne and Sally
Boulder Farm
Hopkinton, NH 03229

Ross, Wayne and Ruth
Rossview Farm
Concord, NH 03303

NEW JERSEY
Post, Gary and John Wolters
Sussex Co. Strawberry Farm
Newton, NJ 07860

Robson, Neil and Jean
Robson Farms & Greenhouse
Wrightstown, NJ 08562

NEW YORK
Green, Norman and Glenn
Green Bros. Apple Hills
Binghamton, NY 13905

Jackson, Skip and Jeanne
Iron Kettle Farm LLC
Candor, NY 13743

Fenton, Joel and Marilyn
Fenton's Berry Farm LLC
Corfu, NY 14036

Chase, Robert and John
Chase Farms
Fairport, NY 14450

Borchert, Marion*
Ernest and Marion Borchert
Borchert Orchards
Marlboro, NY 12542

Schmitt, Teresa
F. & W. Schmitt Bros. Farm
Melville, NY 11747

Behling, William & Marion
Behling Orchards
Mexico, NY 13114

Kent, Helen*
Jim and Helen Kent
Locust Grove Farm
Milton, NY 12547

Wiles, Frank and Ellen
Our Green Acres
Owego, NY 13827

Tomion, Alan
Tomion Farms
Penn Yan, NY 14527

Pearson, Kathy and
Craig Michaloski
Green Acres Farms
Rochester, NY 14612

Secor, Mary*
Donald and Mary Secor
Secor Strawberries Inc.
Wappinger Falls, NY 12590

DeGraff, David and Joan
Davey-Joan's Strawberries
Williamstown, NY 13493

Gifford, Robert and
Katherine Sherras
Oronacah Farm
Clifton Park, NY USA 12065

Joy, Anthony and Rita
Joy's Strawberry Haven
Fredonia, NY 14063

Dressel, Roderick and Ethel
Dressel Farms
New Paltz, NY 12561

NEW YORK *(continued)*
Merkley, Donald and Eleanor
Don-Elea Farms
Ogdensburg, NY 13669

Dykeman, Henry and Jean
The Henry Dykemans Inc.
Pawling, NY 12564

NORTH CAROLINA
Cline, Ira and Ann
Ira Cline Farm
Conover, NC 28613

Lineberger, Debbie*
Ervin and Debbie Lineberger
Killdeer Farm
Kings Mountain, NC 28086

Carrigan, Douglas and Judy
Carrigan Farms
Mooresville, NC 28115

Lineberger, Harold and Patsy
Lineberger's Maple Springs Farm
Dallas, NC 28034

OHIO
Spiegelberg, Dale and Jessie
Spiegelberg Farms
Amherst, OH 44001

Patterson, James and Nancy
Patterson Farms, Inc.
Chesterland, OH 44026

Rhoads, Brent and Kathy
Rhoads Farm Market
Circleville, OH 43113

Ross, Craig and Rita
Red Wagon Farm
Columbia Station, OH 44028

McClay, Charles & Roger Hayes
Hayes and McClay Farm
Groveport, OH 43125

Roediger, Roger and Evelyn
Roediger Berry Farm
Hilliard, OH 43026

Ford, Jonathan and Barbara
Rock Bottom Farm
Middlefield, OH 44062

Fulton, William and Joyce
Fulton Farms
Troy, OH 45373

Waters, Wendell
P.B.F. Farms
West Lafayette, OH 43845

Jackson, Don, Lee, Tom & Karen
Jackson's
Xenia, OH 45385

Rosby, Wm. and Michael
Rosby Bros. Berry Farm
Brooklyn Heights, OH 44131

Jutte, Oscar and Suzanne
Ralph Jutte Farms Inc.
Fort Recovery, OH 45846

PENNSYLVANIA
Seesholtz, John and R. A.
Seesholtz Bros. Inc.
Bloomsburg, PA 17815

Den Hartog, John and Coby
John Den Hartog Orchard
Fayetteville, PA 17222

Barber, James E.
Franklin, PA 16323

Baronner, Robert
Baronner Farms
Hollidaysburg, PA 16648

Funk, Amos and Esta
Funks Farm Market
Millersville, PA 17551

Beinlich, Carolyn*
Ron and Carolyn Beinlich
Triple B Farms
Monongahela, PA 15063

Reilly, Cherie*
Mike and Cherie Reilly
Reilly's Summer Seat Farm
Pittsburg, PA 15237

Elbel, Glenn and Dorothy
Elbel's Produce Farm
Punxsutawney, PA 15767

Finch, Ron
Claron Farms
Waterford, PA 16441

Troyer, Ronald and Debora
Troyer's Strawberry Acres
Waterford, PA 16441

Rotthoff, Walter and Virginia
Rotthoff Farms
Wattsburg, PA 16442

Constable, Stuart
Highland Orchards Inc.
West Chester, PA 19380

Pallman, Richard
Pallman Farms
Clarks Summit, PA 18411

Shenk, John and Linda
Lititz, PA 17543

Irish, Fred and Leonora
Irish Farms
Coudersport, PA 16915

Hellerick, Karl and Doris M.
Hellerick's Farm
Dublin, PA 18917

Sowieralski, Tom and Richard
S-Berry Farm
Frederick, PA 19435

Stutzman, Doris*
Stutzman Farms
Indiana, PA 15701

SOUTH DAKOTA
Gjevre, Marilyn*
Paul and Marilyn Gjevre
Sweetwater Farm
Rosholt, SD 57260-9535

TENNESSEE
Scott, David and Steve Wayne
Scott Farms
Unicoi, TN 37692

TEXAS
Copeland, Don
King's Orchard
Plantersville, TX 77363

VERMONT
Gray, Bob and Kim
Four Corners Farm
South Newbury, VT 05051

Pierson, David
Pierson Farms
Bradford, VT 05033

Winslow, Mark and Andrea
Pittsford, VT 05763

VIRGINIA
Fulks, M. R. and Judy
Belvedere Plantation
Fredericksburg, VA 22408

Bowman, Jim and Cathy
Mountain River Gardens
Grottoes, VA 24441

Conboy, Jim and
Steve Blankenbaker
Rapidan Organic Farm, LLC
Rapidan, VA 22733

Geyer, Charles and Anne
Westmoreland Berry Farm
Oak Grove, VA 22443

Shank, Ray and Marietta
Strawberry Acres
Rochelle, VA 22738

Goode, W. Aaron and Betty
Chesterfield Berry Farm
Moseley, VA 23120

WASHINGTON
Sakuma, G. and S. R. Bryan
Sakuma Bros. Farms Inc.
Burlington, WA 98233

Biringer, Michael and Dianna
Biringer Farm
Marysville, WA 98271

WEST VIRGINIA
Yoder, Linda*
Delmar and Linda Yoder
Owl Creek Farm
Morgantown, WV 26505

WISCONSIN
Statz, Beatrice*
Roman and Bea Statz
Statz's Berryland
Baraboo, WI 53913

Thompson, Jeff and Marcia
Thompson Strawberry Farm Inc.
Bristol, WI 53104

Truettner, Gary and Joan
Truettners Berry Farm
Manitowoc, WI 54220

Kluth, Dale and June
Glendale Farms Inc.
Clintonville, WI 54929

Johnson, Arnold and Gloria
Johnson's Berry Patch
Iron River, WI 54847

Struye, Linda
The Little Farmer
Malone, WI 53049

Scheel, Dan and Karen
The Elegant Farmer Inc.
Mukwonago, WI 53149

Secher, Dale and Cindy
Carandale Farm
Oregon, WI 53575

Lurvey, Mark and Greg
Lurvey Farms
Whitewater, WI 53190

CANADA
(in alphabetical order by province)

ALBERTA
Fedak, Grace and Elvin Saruk
Serviceberry Farms
Strathmore, Alberta,
Canada T1P 1K5

Friesen, Ron and Pauline
Dunvegan Gardens Ltd.
Fairview, Alberta
Canada T0H 1L0

Galloway, Eldon*
Eldon and Jacqueline Galloway
Berried Treasures
Fort Saskatchewan, Alberta,
Canada T8L 2T2

BRITISH COLUMBIA
Truscott, Chuck
C & E Farms
Wynndel, British Columbia,
Canada V0B 2N0

MANITOBA
Wolgemuth, Pat and Camilla
Chanlena Berry Acres
Landmark, Manitoba,
Canada R0A 0X0

NEWFOUNDLAND
Lomond, Paul & Shirley
Lomond Farms Ltd.
Steady Brook, Newfoundland,
Canada A2H 2N2

NOVA SCOTIA
Wenham, Dianne*
Terry and Dianne Wenham
Terridi Farms
Bass River, Nova Scotia,
Canada B0M 1B0

Millen, Curtis and Anne
Millen Farms Ltd.
Great Village, Nova Scotia,
Canada B0M 1L0

Webster, Greg
Webster Farms Ltd.
Cambridge Station, Nova Scotia,
Canada B0P 1G0

Hanna, Laurie and Peggy
D. L. Hanna & Sons Ltd.
Parrsboro, Nova Scotia,
Canada B0M 1S0

Casson, Bela*
Strawberry Growers
Association of Nova Scotia
Truro, Nova Scotia,
Canada B2N 5E8

ONTARIO

Lindley, Peter and Joan
Lindley Farms Ltd.
Ancaster, Ontario,
Canada L9G 3L1

Watson, Ted and Paul
Ted Watson Farms Ltd.
Bowmanville, Ontario,
Canada L1C 3K3

Manley, Eva and Ralph
Fairview Farm
Brampton, Ontario,
Canada L6V 3N2

Pate, Tom and Dawn
Brantwood Farm
Brantford, Ontario,
Canada N3T 5L8

Hughes, Peter and Elizabeth
Salem Farms
Colborne, Ontario,
Canada K0K 1S0

Leeds, Paul and Renita
Leeds Berry Farm
Guelph, Ontario,
Canada N1H 6J2

Dentz, Paul and Calvin
Dentz Orchard & Berry Farm
Iroquois, Ontario,
Canada K0E 1K0

Rondelez, William and Jerome
Black Bear Farms
Kingsville, Ontario,
Canada N9Y 2E6

Whittamore, Dave, Mike & Frank
Whittamore's Berry Farm
Markham, Ontario,
Canada L6B 1A8

Ferguson, Mike and Cheri
Ferguson Produce
St. Thomas, Ontario,
Canada N5P 3T1

Cooper, John and Gary
Strawberry Tyme Farms Inc.
Simcoe, Ontario,
Canada N3Y 4K1

Passafiume, F. J.
Applewood Farm
Stouffville, Ontario,
Canada L4A 7X5

Heeman, Rudy and Florence
Heeman Strawberry Farm
Thorndale, Ontario,
Canada N0M 2P0

Belluz, Don and Claire
Valley Berry Patch
Thunder Bay, Ontario,
Canada P7C 5N5

Parks, Bill and Diane
Parks Blueberries
Bothwell, Ontario
Canada N0P 1C0

Gammond, Gerry & Susanne
Gammond Farm
Thunder Bay, Ontario,
Canada P7C 4V2

Chesney, Robert
Thames River Berries
Innerkip, Ontario,
Canada N0J 1M0

Dekok, Jack and Mary
Dekok Family Berry Farm
Kanata, Ontario,
Canada K2K 1X7

Becker, Murray and Sharon
Becker's Berry Patch
Trout Creek, Ontario,
Canada P0H 2L0

QUEBEC

Charbonneau, Simon
Fraisebec Inc.
Sainte Anne Des, Quebec,
Canada J0N 1H0

SASKATCHEWAN

Epp, Ann*
Ben and Ann Epp
Strawberry Ranch Inc.
Saskatoon, Saskatchewan,
Canada S7K 3J6

FOREIGN
(in alphabetical order by country)

AUSTRALIA

Pettinella, M. L. and J. L.
Coombe Berry Farm
Coldstream, V1C
Australia 3770

Lewis, Rodney A.
Agon Pty. Ltd.
Pooraka, Australia 5095

BELGIUM

Bal, Eddy
Veiling Borgloon C.V.,
Borgloon, Belgium

ENGLAND

Neal, Brian and Wendy
K & S Fumigation Services Ltd.
Tenteeden, Kent
England TN30 71 IT

GERMANY

Haitz, Bernhard and Luzia
Erdbeerparadies H. Koffler
Durmersheim, Germany 76448

ITALY

Martinelli, Alessio
CIV Consorzio Italiano Vivaisti
Ferrara, Italy 44020

THE NETHERLANDS

Sikma, T.
Goossens Flevoplant
Ens, The Netherlands 8307PJ

SPAIN

Aguilar, Ramon
Huelva S.A.-Viveros
Sevilla, Spain

SWEDEN

Rydewald, Leif
Bergums Bio. Barodling
Olofstorp, Sweden S 42491

SWITZERLAND

Wolfensberger, Hansheinrich
Dipl. Ing., Agr. ETH
Pfaeffikon, ZH
Switzerland 8330

Bibliography

Allardice, Pamela.
   Strawberries. San Francisco: Chronicle Books, 1993.

American Health. June 1994.

Anton, Liz and Beth Dooley.
   It's the Berries: Exotic & Common Recipes! Pownal, Vermont: Garden Way Publishing,
   Storey Communications, 1988.

BALL 100% Natural® Fruit Jell™. Alltrista Consumer Products Company marketers of
   Ball Brand Home Canning Products.

Bertelsen, Diane. *The Strawberry Industry*.

Brown, Marvin. "Welcome Speech." NASGA Conference, 1995.

Buckner, Sally. *Strawberry Harvest*. St. Andrews Press, 1986.

Buszek, Beatrice Ross.
   *The Strawberry Connection: Strawberry Cookery with Flavour, Fact and Folklore,
   From Memories, Libraries and Kitchens of Old and New Friends—and Strangers*.
   Halifax, Nova Scotia: Nimbus Publishing Limited, 1984.

California Strawberry Advisory Board.

Carey, Michael.
   "The Thing About Farming," *The Noise the Earth Makes*. Pterodactyl Press.

Casson, Bela. Strawberry Growers of Nova Scotia.

Country Woman.

Farm Wife News.

First. April 26, 1993.

Fitness. May 1995.

"Fitness and Nutrition." *The Minneapolis Times*.

Harris, H. H.
   "The Strawberry." (Reprinted from 1930 Wisconsin Strawberry Field Day Program.)

Home Cooking. May 1995.

The Market Link. Queensland (Australia) Fruit & Vegetable Growers.

McCarthy, Leora M. Fox Lake, Wisconsin.

Michigan Beekeeper's Association.

Old-Fashioned Strawberry Recipes with Historic Notes.
 rev. ed. Nashville, Indiana: Bear Wallow Books, 1980, 1987.

Olsen, Margaret.
 "Strawberries: Consumer Attitudes and Usage."
 Queensland (Australia) Fruit & Vegetable Growers. September 1994.

Parade.

Pitzer, Sara.
 Simply Strawberries: a Cookbook. Prownal, Vermont: Garden Way Publishing, 1985.

Produce Business. March 1994.

Produce Guide. 1993.

Sphere.

Strawberry Growers of Nova Scotia & Nova Scotia Federation of Agriculture.

SURE•JELL Fruit Pectin, SURE•JELL For Lower Sugar Recipes Fruit Pectin,
 CERTO Liquid Fruit Pectin. Kraft Foods, Inc.

Whiteaker, Stafford.
 The Compleat Strawberry: The Strawberry in Art, Poetry, and Song With Herbal Remedies, Beauty Hints, Gardening Tips, and 70 Delectable Recipes.
 New York, New York: Crown Publishers, Inc., 1985.

Women's World.

Yoder, Del.
 "Strawberry Thanksgiving." *Wisdom Keepers: Meeting With Native American Spiritual Elders.*

Index

Some grow on the mountains
and woods, and are wild, but
some are cultivated and are
so odorous that nothing can
be more so.

Estinne de ReHortensi (1535)

Index

Strawberry Toothpaste—
Teeth can benefit from regular brushing with **strawberries**. First, to strengthen the gums, mix a handful of dried, powdered leaves and roots of the wild berry, sprinkle your toothbrush with this and brush your teeth and gums well. It will not damage the enamel nor turn your teeth red. If it is not too much trouble, then simply chew a few berries and let them stay in your mouth for a few minutes. Your lips and mouth will feel fresh and the treatment benefits the gums. Support for this remedy, and for the claims of all the old herbal writers concerning the effects **strawberries** on the teeth and gums comes from modern orthodox dentistry. One American Study *Medical Botany* (1977) says that the roots and leaves used as lotions and gargles, help fasten loose teeth.

Home Remedies, The Compleat Strawberry

Sunburn Care—
For mild sunburn, take a few fresh **strawberries** and rub them over affected areas. Leave the juice on for at least half an hour. Wash it off with warm water.

Home Remedies, The Compleat Strawberry

PRAISE FOR *THE WIFE'S TALE*

"To read Aida Edemariam's *The Wife's Tale* is to savor the life of her grandmother, Yetemegnu. It is a life scented with ginger and garlic, cardamom and basil, which spans emperors, revolutions, invasion, conquest, and liberation. Rather than cataloguing Ethiopia's turbulent modern history, Ms. Edemariam stitches together the fragmentary memories and experiences of a single woman"   —*The Economist*

"Edemariam anchors the book in these mundane rhythms, setting them against a vividly realized landscape. . . . Political turmoil sweeps in like a dream. . . . The book elegantly collapses the distance between the vast and the intimate, showing how history reaches even the most sheltered."   —*The New Yorker*

"*The Wife's Tale* is the extraordinary memoir of a woman who lived through the cataclysmic events that shaped modern Ethiopian history. The narrative, which is lovingly and expertly put together by her granddaughter, is a window into a world that would otherwise be invisible to us."   —Abraham Verghese, author of *Cutting for Stone*

"In this outstanding and unusual memoir, journalist Edemariam traces a century of Ethiopian history through the life—and distinctive voice—of her nonagenarian grandmother, Yetemegnu. . . . It's evident she spent years gathering not just her grandmother's tales but also traveling and researching historical archives in order to bring alive Ethiopia's shift from the age of empires to its present-day, turbulent democracy. But Edemariam doesn't let the scaffolding of her research show. *The Wife's Tale* is told with the turns and twists of a novel, layered with dialogue and stories."   —*Financial Times* (London)

"[Like Chaucer's "The Wife of Bath's Tale"] Aida Edemariam's sublimely crafted tribute to her grandmother also involves sparring storytellers, religion, and the happiness and sovereignty of married women. . . . Intimate history meets the sweep of imperial history when Yetemegnu finds the courage to resist. . . . Yet the role that Yetemegnu finally inhabits is not that of mother but storyteller. . . . Her story is certainly cracked open in the telling, so assured and transcendent, it could win Chaucerian contests."

—Gaiutra Bahadur, *New York Times Book Review*

"This biography of a heroic Ethiopian woman is surely unique, above all for its brilliant combination of big historical vistas with vivid physical details (the clothes, the cooking, the weather!). To begin with I found the narrative almost overwhelming, a sort of sensual overload, but gradually I became completely mesmerized, and found myself carried deep into that other world, with all its beautiful customs and strange cruelties and enduring loyalties. An exceptional biography that really does open a human window onto an unknown history."

—Richard Holmes, author of *Footsteps* and *The Age of Wonder*

"At once a poignant, intimate memoir, a revelatory history, and a formally inventive work in which the truth is made as strange as the best fiction."

—Judges of the 2014 Royal Society of Literature Jerwood Awards for Nonfiction

"In the hands of Aida Edemariam, a strong, poetic writer, a seemingly ordinary life opens up to reveal the extraordinary richness at its heart. . . . The power of Aida Edemariam's writing is its ability to reach across the gaping chasm formed by time, alien tradition and unfamiliar mores, connecting up our common humanity."

—Michela Wrong, *The Spectator* (London)

"A remarkable book. . . . Aida Edemariam, viewing the country's past century through her grandmother's eyes, brings us a nuanced view, informed by a real knowledge. . . . She respects the past, but she doesn't sentimentalize it. To read *The Wife's Tale* is not just to hear about times past and (for a western reader) far away, but to be transported into them." —Lucy Hughes-Hallett, *New Statesmen* (London)

"Extraordinary vivid 'personal history' . . . Edemariam not only brings her grandmother to life but also conveys the complexity of a unique, still strongly religious African culture. . . . She weaves in all the necessary historical detail, while expressing in precise and often lyrical language the colors and textures of a beautiful country and the customs and mind-set of an ancient people." —Andrew Lycett, *Literary Review*

"*The Wife's Tale* is a remarkable achievement: meticulously researched, finely wrought, and deeply felt, it is the story of one woman's life lived, not so much against the backdrop of history, but in the midst of it. Edemariam's grandmother succeeded in building a life out of very little, except her enduring, quiet courage." —Aminatta Forna, author of *The Devil That Danced on Water*

"[Yetemegnu] emerges as a bewitching and resilient figure whose life-changing moments sometimes intersect with the tumultuous history of her nation. . . . What brings this narrative flaring to life, though, is not the rigor of its research but its imagination and novelistic tone; Edemariam's prose climbs inside Yetemegnu's memories to inhabit them and bring her solidly, vividly, to life. . . . The physical world . . . is invoked with a richness that feels tangible, sensuous." —Arifa Akbar, *The Observer* (London)

"This account of the life of Aida Edemariam's grandmother is embellished with the author's fiery imagination and her deep reading about Ethiopia's history. . . . It's a book that gets under the skin. —*The Times* (London)

THE WIFE'S TALE

AIDA EDEMARIAM

THE WIFE'S TALE

a personal history

HARPER PERENNIAL

NEW YORK • LONDON • TORONTO • SYDNEY • NEW DELHI • AUCKLAND

Originally published in Great Britain in 2018 by Fourth Estate.

FIRST HARPER PERENNIAL EDITION PUBLISHED 2019.

Library of Congress Cataloging-in-Publication Data has been applied for.

ISBN 978-0-06-213605-3 (pbk.)

19 20 21 22 23   OFF/LSC   10 9 8 7 6 5 4 3 2 1

For Rahel

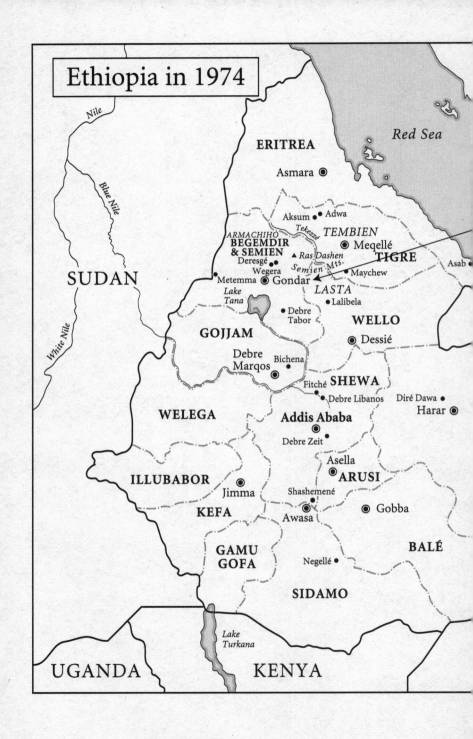

Ethiopia in 1974

Nile

Blue Nile

White Nile

Red Sea

SUDAN

ERITREA

Asmara ◉

Aksum • Adwa

Tekezé

TEMBIEN

◉ Meqellé

ARMACHIHO
**BEGEMDIR
& SEMIEN**
Deresgé •
▲ *Ras Dashen*
Wegera • *Semien Mts.*
Metemma • ◉ Gondar
Maychew •

TIGRE

Asab •

Lake
Tana

LASTA
• Lalibela

GOJJAM

• Debre
Tabor

WELLO

◉ • Dessié

Debre
Marqos
◉ • Bichena

Fitché • **SHEWA**

• Debre Libanos

Diré Dawa •
Harar ◉

WELEGA

Addis Ababa

• Debre Zeit

ILLUBABOR

• Jimma

Asella
◉ **ARUSI**

Shashemené
◉

KEFA

Awasa
◉

◉ • Gobba

**GAMU
GOFA**

• Negellé

BALÉ

SIDAMO

Lake
Turkana

UGANDA

KENYA

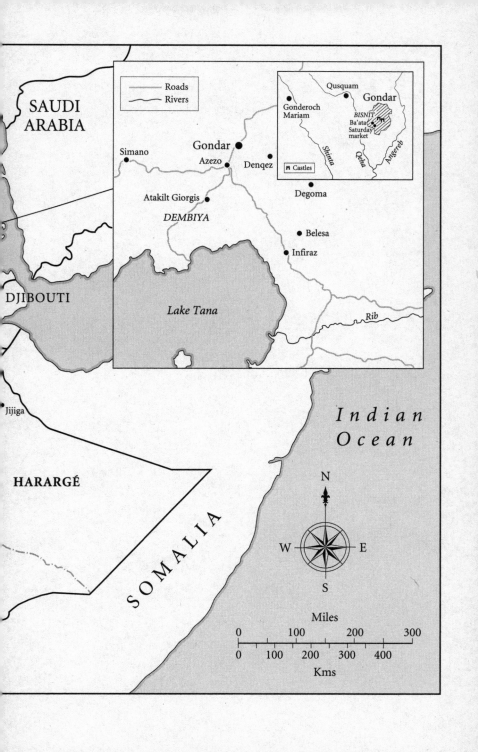

SAUDI
ARABIA

Simano

Gondar

Azezo

Denqez

Atakilt Giorgis

DEMBIYA

DJIBOUTI

Lake Tana

Degoma

Belesa

Infiraz

Rib

Roads
Rivers

Qusquam

Gonderoch
Mariam

Gondar

BISNIT

Ba'ata
Saturday
market

Shinta

Qeha

Angereb

Castles

Jijiga

HARARGÉ

Indian
Ocean

SOMALIA

N

W   E

S

Miles

0   100   200   300

0   100   200   300   400

Kms

PAGUMÉ
THE THIRTEENTH MONTH

*Rains broken by occasional sunshine. Examination of
boys in church school to decide who will be deacons.
End of fiscal year. New Year's Eve.*

Four coals huddled into a low clay pot, glowing red through their
films of ash. My grandmother reached in among the folds of her
shawl and drew from a small pouch a kernel of frankincense. She
dropped it among the coals and at once it melted, hissing, releasing sweet smoke that rose and tangled with the smell of roasting
coffee, of rain gathering beyond the open door, of unfurling earth.

If it rains on Ruphael's Day, my grandmother said, the water is
holy. When we were children we'd tear off our clothes and dance
through it singing. And if there was a rainbow it was as though
Mary's sash had been thrown across the sky.

Above our heads, on the corrugated-iron roof, the rain began.
Thud. Thud. Thud-thud. Each drop carrying with it a sense of
great chill distances travelled, of interrupted speed.

And all through Pagumé anyone young went down to the rivers
before dawn, said my grandmother. You had to get to the water
before the birds could taste it. She held the round-bellied pot high,

products of colonialism?

so the coffee clattered into the little porcelain cups. Added sugar, or salt, or tiny tear-shaped leaves of rue, passed the cups around. I've never liked rivers, though, nor lakes, she said, not since I was a small child.

But even though I was afraid I begged to be allowed to go. I was staying with my grandmother. She was kinder than my aunt, especially when I wet the bed. She'd just turn the jendi over, change the bedclothes. She was patient with me, and loving. Like my mother – and at once my own grandmother was crying, tears spilling into her shawl.

Ayzosh, Nannyé, I said. Ayzosh. Take heart. Yibejish, lijé, she answered. Yes, child, may you be saved. Ayzosh. Yibejish, wiping the wet away. I miss my mother, she said. I know, I answered, I know. So what happened at the river? Steering her back, to distract her as much as anything. Pushing her on, as I did more and more often, knowing many of the stories, but knowing also that there were more, told and retold for decades, shaped, reshaped – or sometimes, when enough time had passed – cracked open in the telling. What did you say? How did you feel, and what do you feel, now?

Sometimes the answers were immediate. Well, I said this, of course, or no, I don't remember the date, or the time, only that the feast of St John was approaching, and I had so much work to do. Or not now, or I've told you that before – though often you could tell it was a rote demurral, that she wanted to continue. Other times the reply was a small smile and a twist into shyness, no, no, those things are not spoken of. When were you happy? I asked once. I'm never happy, came the answer, I'm always crying. All of my life is painted in tears.

The third round of coffee had been drunk, the dregs slopped out into the yard. The smoke drifted into the corners and disappeared. Nannyé held out her hands, palms heavenward. May He bring justice to the wronged, to the poor, to the oppressed. May He clothe the naked and liberate the <u>crucified</u>. May He protect us, and bless us.

I dipped my head. <u>Amen.</u> We watched as sunlight flared through the steam rising from the wet ground, and through the open door. Birds sang.

At last I was allowed to go, she said. We left our houses excited, in the dark, and walked down into the valley. The Qeha had been filling all the rainy season, it moved fast and deep. The other children took off their clothes and jumped in. They cupped the water in their hands and threw it high. They laughed and splashed and wrestled. I edged forward. The water crept toward my toes. I started to move forward again, but I couldn't bear it. I screamed. And I ran.

She laughed, a laugh that took her over as utterly as her tears had a moment earlier. A complicated laugh, deep and delighted but serious also, for in fact she was still afraid and always would be; because she remembered the child she had been so clearly; because in many ways she was still that child.

BOOK I

1916–1930

MESKEREM
THE FIRST MONTH

Floods recede. Yellow masqal daisies cover the land.
New and fallow fields ploughed for cultivation.

AND WHEN THE MAIDEN WAS THREE YEARS OLD IYAKEM
CALLED HIS PURE, HEBREW MAIDSERVANTS AND PUT
CANDLESTICKS WITH WAX CANDLES IN THEIR HANDS,
AND THEY WALKED BEFORE THE MAIDEN AND BROUGHT
HER INTO THE HOUSE OF THE SANCTUARY ... THEN THE
PRIESTS TOOK HER, AND ESTABLISHED HER IN THE
THIRD STOREY OF THE HOUSE OF THE SANCTUARY ...
AND HER KINSFOLK AND THE PEOPLE OF HER
HOUSEHOLD TURNED AND WENT BACK TO THEIR
HOUSES IN GREAT JOYFULNESS, AND THEY PRAISED THE
LORD GOD, AND GAVE THANKS UNTO HIM BECAUSE SHE
HAD NOT TURNED BACK ... AND MARY DWELT IN THE
HOUSE OF THE SANCTUARY OF GOD LIKE A PURE DOVE,
AND THE ANGEL OF THE LORD BROUGHT FOOD DOWN
FOR HER AT ALL TIMES.

– LEGENDS OF OUR LADY MARY THE PERPETUAL VIRGIN
AND HER MOTHER HANNA

By the time the attention turned to her, she was in an agony of restlessness. She had tried to concentrate, to follow the familiar shapes of words she did not expect to understand, to feel their practised roll and pitch, to distinguish between the voices, now muttering, now confident and clear. She had tried to stand still; the effort made her aware of each limb, each finger and toe, of her head balanced on her neck, of the netela, so fine it was near weightless, that covered her head like a cowl. If she moved it gave off a faint scent, of sunshine and new-spun cotton, a wide, outside smell that cut across the eddying incense like an opened window.

She wished she was out there now, playing. Sitting on her haunches to throw a smooth round stone into the air, using the same hand to pick up more stones, then intercept the first stone's descent. Or games that went on and on, till bats swooped and looped through the dusk. Coo-coo-loo! the other children would call, speeding to hiding places. Not yet! she would call back, from her perch on a pile of rocks. Coo-coo-loo! Not yet! Coo-coo-loo – Now! And they would race toward her, vying to touch her skirt and claim themselves safe, making her laugh and laugh. A far better feeling than the time she had ripped up a perfectly good dress to make herself a doll, thinking to strap it to her back as if it were a child. Oh, the whipping she had got then! And the doll had felt too light, lacking the heft of a real baby. It was more fun to play mother with the neighbourhood children. Or weddings, wrapping dolls in scraps of red and green silk and walking them bandy-legged to church.

She shifted, stood still again. The long black cape was lined, the gold filigree around the collar and down the front made it heavy, and it was getting heavier. She hugged herself tight, underneath it. Her stomach was so empty.

The wall of clergy changed position. A book was opened, one wave-edged vellum page at a time. A pause, and a priest looked at

her. At once she looked down. Bare toes on a faded, fraying carpet.
Hers, theirs. So many of theirs.

Repeat after me. If he is ill – if he is ill. The fact of her voice loud
to her. Her breath warm tendrils moving across dry lips, dust
swirled along the ground by an afternoon breeze. If he grows thin
– if he grows thin. Or darkens – or darkens. If he suffers – if he
suffers. Or is in trouble – or is in trouble. If he becomes poor – if
he becomes poor. Even if he dies – even if he dies. I will not betray
him – I will not betray him. A turn away from her, and another
voice, a man's. If she is ill – if she is ill. If she grows thin – if she
grows thin. Or darkens – or darkens. The priest took her right
hand and placed it on his cross. Then he took another hand and
placed it on hers. I will not betray her.

A ring was threaded onto her third finger, another onto
the man's. It would be years before she understood what she
had promised. For the moment all she knew was a thickening
of the air, a seriousness, a flutter of – what? Apprehension,
perhaps.

More prayers. A prayer for the rings, and a prayer over their
capes. A thumb slick with holy oil tracing a rough cross onto her
forehead, and a prayer over that. Hands bearing cushions, and on
the cushions crowns, high straight-sided traceries of gold. A priest
held one aloft for a long moment, then settled it on her head. She
stepped back under the weight. Felt the figure next to her receive
the weight too. The prayer of the crowns, and only then the church
service.

After the bread and the raisin wine, taken under a tilting roof
of heavy brocade; after they had bowed to kiss the threshold of the
holy of holies; after they had walked slowly around it, once, the
priest extended his cross for them to kiss. It was cold, and smelled
of earth after rain.

Ililililil! cried the women.

The sun had burned the mist out of the cedars and hurt her eyes, so she had to use her feet to search for the steps of the low, humped building.

Ililililil!

Out here the trilling was thin, echo-less. Cockerels crowed, and crows answered. Kwaa. Kwaa.

Ililililil!

The congregation assembled at the bottom of the steps and began a slow procession around the churchyard. Past the bethle-hem, with its protective ring of dark evergreens, its nuns picking through baskets of wheat for the eucharist bread; past a young olive tree, leaves quivering silver. A long, stately walk around a central absence: the foundations of the main church were partly covered over with vines and moss, partly naked, as though they had been exposed yesterday. When the circuit was over the congregation settled under trees to listen to the sermon, and to praise-couplets composed for this day. Then, finally, 'May He bless you. May He multiply your seed as the stars in the sky, as the sand of the sea. May He make your house rich as the house of Abraham.'

Ililililil! Ililililil!

As they picked their way out of the gate and started down the road she noticed that the streets and alleyways, usually so busy, were silent, that doors were shut tight. Wobbles of woodsmoke, the odd dog foraging among the stones and bones, roosters crow-ing as always, but otherwise an unnatural hush.

She began to see the holes – ragged holes, punched through sturdy mud walls – and to glimpse the homes inside: raised wooden beds strung with leather, pots and pans, dividing curtains. Once she saw directly through to a front door, barricaded against the disease until the house's inhabitants could fashion their escape. The women noticed her looking. They drew the netela further about her face, and hurried her on.

And then the feasting began. She knew – because she had helped, or been told to run off and play because she was getting in the way – that the women had been cooking for weeks. She had watched the huge earthenware gans of grain in the storehouse deplete, and those of mead and beer multiply, had watched the pounding, the chopping, the sifting, the kneading, had stared as shouting men whipped and dragged five bullocks through the narrow gate. The blood had dried into dark tributaries around the stones in the yard, and now in a corner a dog gnawed at a horned skull.

She was used to eating separately from the adults, to being silent unless spoken to. Silent she was still, but in a confusion of pride and worry. Here was all the attention she had ever wanted – but in such an inversion of her usual state! Everyone made a fuss of her, kissed her, hugged her; even her aunt coaxed her to take sips of mead or, collecting together a little heap of the best pieces of meat, the whitest injera, fed her. She opened her mouth politely, tried not to gag.

Poems again, more joy-cries. Someone beat a drum and was instantly shushed. At this her whole body rose in protest. She thrilled to drums, to music; hearing even the most distant party would slip down the lanes to join in. Why could she not do this now the drummers were in her own home? Her mother noticed. 'My heart, please understand. It draws attention. If we play the kebero, if we dance, the evil eye will notice us and the disease will come here. It's killing people. Remember that lady from the market? She said her waist ached, she had a headache, she rattled with fever. She died yesterday. We cannot risk that. Please understand, child.'

She would always remember no one danced at her wedding. And for the rest of her life she would try to make up for it, threading her way into the centre of the room, placing her hands on her

hips, crooking her neck and – especially after her husband died
– showing everyone how it ought to be done.

The next morning she was given a new underdress. Then another,
for warmth and volume. The main dress was a mass of soft white
muslin edged in red. A necklace, corded black silk wound round
with delicate gold chains, so long on her eight-year-old body that
its two stubby gold crowns swung well below her waist. Silver
anklets. A wide, light netela, draped generous around her shoulders and chest, up over her head, then around her shoulders again
to secure it.

'Nigisté,' said her mother. 'My queen.'

A scatter of hooves, footsteps, a tumble of voices, and then one
of the groomsmen, a relative of theirs, bowing in through the
door, bowing to the women. How did you spend the night? Well,
the women answered, thanks be to God, and you? Well, well, may
His name be praised, may He be thanked for bringing us this day,
may His honour and glory increase, and she was lifted up, up
through a welter of hugs and kisses, prayers and instruction, into
the brightness outside.

The elders were waiting. Past the women, first. Past her smiling
grandmother, her aunt. Then the men. May you be given a long
life. May He watch over you and keep you, rain blessings down
upon you. Her father kissed her. May He go with you, child, all the
days of your life.

Gently the groomsman placed her on a waiting mule. Then,
because she was too young to control it, he mounted too, and,
passing an arm around her waist, grasped the reins. Firmly he
pulled the animal's head round; slowly they moved out of the
compound, and left her family behind.

At first she concentrated on their mount, on the animal's rough narrow back, the part in its mane a dark bolt of lightning. The balls of red wool sewn to bit and bridle that shook at every step. The embroidered saddlecloth. The side of its face, unfeasibly long lashes blinking away flies. The uneven rocking as it searched a path through the stony streets.

After a while she became aware of the running children, the women on errands, the yodelling calls of door-to-door salesmen, the compounds whose walls reflected the sound of their mule's hooves back to them. And the other hooves, too, clopping out a ragged counterpoint. She knew what they carried: narrow embroidered dresses she could wear now; big square dresses, for when she was older; a length of perfectly white, perfectly even cotton; delicate shemmas; thicker shawls edged with wide red bands; fine basketwork woven by cousins over the long rainy season; twelve grey-tinged salt amolés in lieu of silver; gans of dark beer; a case of cured goat hide, just the right shape for a psalter. The slave girl, Wulé, walked alongside them. Another groomsman. And him.

They were led not to the wide das, where, under a temporary roof of saplings and branches, the wedding guests had already gathered, but to the bridal hut nearby. She felt him sit, felt the groomsmen take their places, took her own.

Ilililililililil! Ilililililililil! She recognised none of the women, but the sound was the same.

The noise from the das rose and rose. Rushes of music, a drum – there was no illness in this part of town. Every so often men came to the door, carrying fluttering chickens as offerings to be made into stew for the bridal party; women with fistfuls of pancake and butter, rich food they held direct to her mouth. But she was not hungry. And she still could not trust that she would not wet her bed at night. So she shook her head and refused it all.

The second day. In the das a minstrel sawed at his jasmine-wood masinqo and tossed rhymes like spears into the crowd. Guffaws of recognition only underscored her distance from home. She heard clapping, ililta. They were dancing! Tears dropped onto her hands. The groomsman who had come to fetch her from her family noticed. He caught her small feet in his hands and drew a finger over her anklets. Who bought you these? Do you know how to clean silver? She shook her head.

The third day. Her mouth stuck shut with thirst but she took only the tiniest sips of water, to loosen it. In the das they danced and sang and clapped and cheered but in the little hut, where if she had been old enough the marriage would have been consummated, no one spoke. Then, 'Listen. Have you heard the story of the hawk and the tortoise?' No. 'Shall I tell you?' A slight nod. For the rest of the day the groomsman dredged his memory: The monkey king. The tortoise and the hare.

The fourth day. 'A cat and a mouse were getting married. On the day of the wedding the cat's groomsmen gathered and together they made for the bride's house, dancing and singing in anticipation of a feast. The mouse, like all brides, waited amongst her kinsfolk to be taken away to her new life. Then one of the other mice piped up – "You know, cats can't be trusted. Let's dig holes in the ground, just in case." So they set to it, scrabbling out deep tunnels with hidden entrances. Finally the cats came into view, chanting "Ho – pick one up here, ho, pick one up there, ho." When the mice saw them approaching they turned as one and plopped into their holes. And the cats, who thought they'd been so clever, didn't catch a single mouse.' His laugh wilted into the silence.

The fifth day. 'Aleqa Gebrè-hanna, the famous wit – he was also leader of the church in which you were just married, did you know that? – was walking along the road when he met a donkey-driver. He greeted the peasant with unusual politeness for someone of his high status, even bowing low. In the mead-house later that evening the donkey-driver regaled his friends with his tale of a grand personage who had deigned to speak to him so kindly. But his friends were sharper than he was, and asked for the scholar's precise words. "How are ye?" he repeated, realising, as he did so, that he had been included with his donkeys.'

She laughed. Her head tipped back, her veil slipped off, and for the first time she saw properly the man who had sat there all along. Pure white jodhpurs, wound tight around his calves. A wide sash around his waist. A cape of thick black wool falling from thin shoulders in generous folds. And under the white turban a small dark face and a tiny, straight nose. Awiy! she said in a low voice to the groomsman next to her. When I have children they're going to look like that! He laughed, but she was serious. She dragged the material back over her face.

When, after nearly two weeks, the feasting was finally over, the little party left the bridal hut and walked into the das. It smelled of incense, of food and stale beer. The reeds and wildflowers that had been strewn across the ground were bruised and limp. Even under her netela she felt the expectant eyes; when it was lifted away so the guests could see her it was like a blinding.

AND HE TOOK THE MAIDEN TO HIMSELF, AND HE SAID UNTO HER, 'BEHOLD, O MARY, I HAVE TAKEN THEE FROM THE HOUSE OF THE SANCTUARY OF GOD, BUT I WISH TO GO ON A JOURNEY. TAKE CARE OF THYSELF UNTIL I RETURN TO THEE, AND I WILL ASK THE LORD GOD TO PROTECT THEE AND BE WITH THEE.'

– *LEGENDS*

tree was a green cave, full of shifting underwater light. And so quiet. She drew bare soles along the rough branch and resettled her spine against the trunk. In a minute she would climb down, hugging a bounty of peaches in the lap of her dress. But in a minute. First she wanted just to sit here, in the bird-sewn silence.

When, inevitably, she heard her name called, she didn't answer. Maybe if she was really still – but the calls came closer, till they were beneath her feet. 'Come down. Please come down?' A pause. 'You're the wife of a big man now. You *must* come down.'

But he's *away*, she replied fiercely, though only to herself. And I wish he'd never come back.

'That's right, careful.' She shrugged away the proffered hand, and made for the house.

They were preparing for a visit from her father. Her aunt, Tirunesh, presided. Woizero Tirunesh, in her layers of white shawls, sitting stately, giving orders. Woizero Tirunesh, with her ever-present horn of dark beer, stirring up yet another domestic storm.

She was ambivalent about her father, Tirunesh's younger brother. Mekonnen Yilma was proud, tall, a fast walker, a fast talker, a natural soldier. A good storyteller, too, and committed to witty conversation. Listening to the thrust and flash of his talk, his quick laughter and tight puns, or watching him settle himself onto a stool, take a sip of mead, then lean into his high ten-string lyre to sing slow low Lenten songs, she could almost forget she was afraid of him; almost forget the terror with which she watched him punish the other children, the spell of the thin hide whip curving through the air and kissing the backs of their legs.

Her father was proud, yes, but not too proud to beg. Over the years, from story after story – not all his – she had pieced together how she came to be. When Setechign – pale, beautiful, much-sought-after – first married Mekonnen the expected children did not arrive, so Tirunesh had taken herself off to church to have words with God about the situation. A boy was duly born, and named Nega, for the dawn.

Then Setechign left. No matter that Mekonnen was now a customs officer on the long lawless border with the Sudan. No matter that his immediate master was married to the empress's sister. No matter that he regularly returned from military skirmishes laden with trophies and prisoners of war who often became valuable slaves, and that to this was added the tithes of the peasants who farmed his lands. Nor did it matter that – as Mekonnen, fond of genealogy (particularly his own) and possessed of a preternatural memory for names, made a point of reminding her – he could claim descent from at least three emperors and an empress, Taitu, and would thus give her offspring royal blood. No. His family was too big, there were too many hangers-on, it was all too much. So Setechign went.

He pleaded. He sent emissaries: his sister, his mother, elder after elder. They applied all the subtle pressures of home and hearth, and it took a couple of years, but eventually she returned, and they conceived a daughter.

In gratefulness Mekonnen plied his wife with gifts: trains of donkeys laden with wheat, barley, the whitest teff; pots of spiced butter, baskets of deep-red dried chillis, of cardamom, frank-incense and rue.

Their daughter arrived the day before Christmas, on the feast of Ammanuel. There was no disagreement about what her baptismal name ought to be – Weletè Amanuel, daughter of Amanuel – but her daily name was another issue altogether. Her maternal grandmother favoured Genet, for the garden of Eden; her paternal grandmother Gedamenesh, or my sanctuary; while her mother called her Nigisté, my queen. Tirunesh and her father, however, chose Yetemegnu, or 'those who believe', and it was they who prevailed.

Again Setechign left. Distressed and in spiritual need she walked to Infiraz, where she had heard there was a great zar doctor. Perhaps he could heal her. But when she met him she was afraid. A huge, powerful man, he had once, when he was nineteen, crossed the high Simien on foot, scrambling down gorges and tramping over plains looking for his own mother, who had been abducted by a Tigréan lord. He found her, but could not release her, and on his way back had become possessed by a spirit that had never left him. He had learned instead to control it, and now his home was filled with incense and the smell of roasting coffee, with women speaking in tongues or stamping out their individual spirit dances; dancing, often, to his personal bidding too.

And this particular woman, with her tight-braided hair and neck so long it could take seven rings of tattoos – the zar doctor liked this particular woman. And the more Setechign fought him, the more she hated him, the more he liked her. She hated him so much that when she became pregnant with his child she tried to kill it, standing for hours under a waterfall in the hope of dislodging the growing thing. But it would not leave before it was ready and when it was born she called the boy 'imbi alè', or 'he said no'. Not until he went away to school was his name changed, to Gebrè-Selassie.

Only Setechign's third union, to a rich trader, gave her a measure of calm. The home she made with him was a place of dancing and honey wine, where an animal was slaughtered nearly every week, and parties included the neighbourhood poor, invited to eat their fill. Of warmth and love, where special meals were cooked for a shy daughter who basked in the unaccustomed glow of feeling singular, precious; who had gently to be encouraged to eat and was seldom allowed to stay. Yetemegnu made little protest, but every time she was taken away the grief curled deeper into her heart.

At her aunt's there were fewer parties. Tirunesh was pious and severe, a disciplinarian who had little patience for a child who lost herself in games and dreams. But she made sure Yetemegnu learned to spin, and to cook. Every morning the child was required in the kitchen, to watch and to learn, and at last to try a few things herself – to feel the exact point at which it was best to add spices to onions and garlic turning gold in the pot; to judge just how thin to make the sourdough, so the injera would be delicate and light.

Their work was accompanied by a drone of words, but Yetemegnu never listened in any conscious way. Nor did she take much notice of the slight deacon who appeared after church each day, read from the homilies of Ruphael or of Mikael, then, having been fed, slipped away again, while the women turned to drinking coffee and pronouncing on the generally disappointing ways of the world.

Tirunesh had been watching the deacon, however. She questioned him about his education, his ambitions, and liked his considered answers. 'If I had a daughter,' she said to her husband one day, 'I would marry her to this man.'

So when the deacon's patron, a friend of hers, approached her with the news that the deacon was nearing the end of his training and looking for a wife, the suggestion that this wife be Yetemegnu fell on receptive ears.

'He's just another student from God knows where,' said Mekonnen, disgusted. 'Able, maybe, but so what? There are hundreds like him, cluttering up the churches. Absolutely not.'

'He would be a good husband, perhaps the best she could have.' To Mekonnen this was patently untrue. He could not believe his own sister could so easily squander their lineage on a nonentity; a nonentity, moreover, from Gojjam, an entirely different province, and thus foreign. 'We don't know anything about him. We don't even know who his father is. No.'

'Setechign,' said Tirunesh on a visit one day, as gently as she knew how. 'Isn't it time your daughter was betrothed? The deacon who reads to me –'

'She's a child. She's barely eight years old. I will not give my daughter to a man of thirty who has no women in his household, no mother in evidence, no nurse to care for her. How can you think of such a thing?'

Tirunesh turned to the elders. Deputations arrived at Mekonnen's house, bearing blandishments, arguments, testimonies of character. Mekonnen listened, resisted all of them.

'Look at me!' cried his sister. 'Look at me! I'm barren. Is that what you want for her? I'll curse you for your cruelty!'

'Now, now, no need –'

But she would not hear. 'If you do not marry her to this man I will hate you forever. As Mary is my witness I will never visit your graveside. And you will never stand at mine.'

It was the strongest threat in the armoury, and her brother acceded with an angry sigh. 'Very well. She can marry the student.'

His relenting made it harder for Setechign to hold fast. And different arguments were used with her. Of course the girl was young, but that was common and had its advantages: she could be moulded to her husband's ways, she would grow up in an educated, pious house. It would be good for her. As for the lack of nurturing women, a nurse could be hired, servants, she could be given an experienced female slave.

No one told Yetemegnu what had been decided. Why would anyone bother to tell a girl child?

COME, O JEREMIAH, AND MAKE A LAMENTATION FOR MY
MOTHER HANNA, FOR SHE HATH FORSAKEN ME, AND I AM
ALONE IN THE HOUSE OF BRASS. WHO WILL POUR WATER
ON MY HANDS? AND THE TEARS START IN MY EYES.

– *LEGENDS*

Before her husband left for Addis Ababa to petition for a parish,
he had gone to see the governor of Gondar, to tell him of his
marriage, and through his marriage, of his promotion both in
Gondar society and from deacon to priest; to tell him that his wife
was young and he a man who owned nothing. The governor had
responded as the new priest hoped, awarding him a salary of
twelve quintals of grain, teff and barley, a quintal of chillis, a
generous measure of butter. Every month these things arrived on
donkey-back and were received by her maternal grandmother,
into whose care she had been returned. Here, too, they sometimes
reminded her to put away childish things, but her life was really
not so different from that of other children her age. She settled in
quickly, helping around the house, visiting neighbours, family
friends, sometimes forgetting, for hours at a time, that she could
not stay. Now she really learned to dance, watching women at the
weddings of her grandmother's friends and relations, then slipping in among them and echoing every move. And when they saw
how she loved it, how well and naturally it came to her, they
circled her, and clapped and trilled and sang, encouraging her,
laughing as she responded with tighter, more demanding movements, improving from day to day until she was nearly as good as
some of the more accomplished adults.

One day she was passing the receiving room when she saw her
grandmother had a visitor. This was nothing more than routine
– incense, roast chickpeas, coffee, questions, how are you, and

how are you, and well, thanks be to God. But something made her hesitate in the doorway. The visitor looked at her and set down her cup. 'Your mother is tiring. You must come at once.'

Setechign had been ailing for months. On Yetemegnu's last few visits she had sat by her mother's bedside, trying to manage the fear that rose through her body when her mother complained of what felt like knives cutting through her stomach and refused to eat. Then, three weeks ago, her brother Nega had taken a month's supply of food to their father Mekonnen, who, having come off worst in a dispute with a rival, was imprisoned in Debrè Tabor. On the way home the boy had had to swim across a river; that night a fever clenched his teeth and threw his head back in a rictus of pain. Holy water, administered in dousings and drenchings and trickles through rigid jaws, did not help, and he died the next midnight. The governor took pity on his prisoner and released Mekonnen so he could mourn his son. For three days Mekonnen had sat with his lyre, weeping, singing of his beautiful swimming boy. Setechign simply weakened, disappearing further and further into the hollow under the blankets.

Her mother's house was full of people when Yetemegnu arrived. They cried in the corners and wept through the receiving rooms. They held her, and led her into the bedroom.

Setechign had already been washed and laid out. Her big toes had been tied together, as was done for Lazarus, and her thumbs, so her arms ended in a spear-point aimed at her feet. She had been wrapped in a winding sheet of rough white cotton, and then in a palm shroud. The child drew near, and stood by her mother's head. The hair was glossy, the eyes closed. They would not open – decades later Yetemegnu would remember and weep as if it had just happened.

After the short service, and the first prayer for absolution, scores of people – priests and deacons, relatives and neighbours

who had eaten Setechign's injera and drunk her mead – followed
the bier out of the house and down the road toward the church.
The bearers had not travelled far, pacing slow, leading a low
hubbub of gossip and care, when they set her down. At once the
chat stopped, and the crying began again, the women leading. The
deaths Setechign had suffered, the lives she had brought into
being. Her loves, her lineage, her generosity, called out, rhymed
out, echoed in chorus. Then the deacons sang another prayer, the
bier was lifted, and they carried on. Seven times, so all the
thoroughfare knew of her passing.

In the churchyard she was set down while her male relations
dug into the ground. A smell rose, of loam and of rain. Yetemegnu
was brought to the front. Now she could see the priest who clam-
bered into the shallow grave; see his censer swinging, one corner,
another, another, overlaying earth with pious perfume. Hear the
final prayers. Watch the bending backs lower their freight into the
ground, head to the east, feet to the west, feel, like a blow to her
own body, the first handful of soil land upon her mother.

In the waning years of the Gondarine age, when emperors became
puppets and warlords danced them on and off their thrones as
mood and circumstance took them, Emperor Teklè-Haimanot II,
godly, handsome (and not a little vain), tried to live up to his name
by planting seeds of piety wherever he went. By the end of the
eighteenth century, when he was ushered into a monastery by a
brother eager to take his turn as puppet-in-chief, he had established six churches, among them a structure he at first called
Debrè-hail-wa-debrè-tebab, mount of might and mount of
wisdom, and then, because it was consecrated on the feast of
Mary's Presentation to the Temple, Ba'ata Mariam.

Ba'ata was, from the beginning, well endowed. Teklè-Haimanot settled upon it fertile lands that stretched down into the Bisnit and Qeha valleys, into Gabriel, and even to the districts of Dembiya and Deresgé, a whole day's journey away – lands from which a fifth of all harvests flowed back to the church. A spring was discovered and designated holy. Ba'ata's tabot, its life-giving replica of the Ark of the Covenant, was of marble, and the emperor commissioned the best of fresco-painters to illuminate its walls. By the early 1800s Ba'ata was among the richest, most powerful, and, some said, most beautiful of the forty-four churches in Gondar. Students walked for days to study under its dark trees, learning the syllabary, the psalms, the homilies of Mary, and especially the aquaquam, the slow dance of David before the Ark, of which Ba'ata claimed 276 masters.

When, some fifty years later, Emperor Tewodros II's chronicler described the capital's priests as debauched occultists (and his liege, of course, as the opposite of these things), there was perhaps something in it. Certainly they were not accustomed to being gainsaid, and especially not by a brawling upstart they mocked for being born to a mother so poor she'd had to sell purgative kosso to survive; so poor, one story went, the priests of Ba'ata turned her away when she brought her son to be baptised: she could not afford the two jars of dark beer, two bowls of stew and forty injera they demanded in payment.

But they would have done well to remember that this so-called upstart had also defeated lord after warlord to become emperor in act as well as in name, because they soon found that the churches, with their vast tracts of land and internecine theological disputes, were next. Five years later Tewodros stripped Gondar of its status as capital; ten years after that he seized from its churches any land he deemed surplus to requirements; finally, on the sixth day of the third month, when, wrote his chronicler, the very stars 'began to fly about as though struck by fear', Tewodros sacked the sanctuaries and set fire to the city. Castles, homes, churches – everything burned. Bells, chalices, drums, censers, crosses, manuscripts were torn out of their places and taken for his treasuries. Priests who fought to keep them were fed to the pyres.

In the silence after Tewodros and his soldiers were gone, as embers flickered against the dark like so many more burning towns, Ba'ata counted its blessings. The grand outer circles, the frescoes and the holy of holies smouldered and smoked, but the tabot, being marble, had not been consumed, and so the heart of the church was intact. The vestments encrusted in gold and silver, the sistra and the drums, the illuminated manuscripts, had been hidden underground, in a chamber below the holy of holies, and they too had survived. The priests built a temporary hut in the grounds and continued their ministry.

But they were again besieged. 'Oh master!' they wrote to Tewodros's successor, Emperor Yohannes IV, borrowing from Psalm 79, 'The heathen have come into thine inheritance; thy holy temple have they defiled; they have laid Gondar in heaps.' But Yohannes was already occupied, leading an eighty-thousand-strong force against Italian armies threatening to take the Eritrean highlands, so he asked Menelik, king of Shewa (and his chief rival) to intercept the Sudanese Mahdists advancing on Gondar. Menelik did not arrive in time: the jihadists razed nearly every remaining church to the ground. Only two escaped – Medhané-Alem and Debrè-Birhan Selassie, the latter protected, people said, by a swarm of holy bees.

*

By the early years of the twentieth century, when Tsega first followed his teacher of scriptures through the fields and thick woods, Gondar, which at its zenith had held up to seven thousand souls, was home to less than a tenth of that number. The castles, once hung with silk and ivory, chalcedony and Venetian glass, were bare and cold, fluttering with bats and pigeons. Thatched huts huddled as if for warmth against the outer walls. Only on Saturday, market day, did the town manage to summon up something of its former bustle.

Though Ba'ata, just up from the main market, had suffered an inevitable winnowing of its congregation, the itinerant students came still, and Tsega joined them. In the little village in Gojjam where he was born he had gone to church school with all the other wide-eyed boys, learning his alphabet in sing-song call and response. He had learned to write, shaping his letters so they fitted onto the bleached shoulderblades of sheep, because these were plentiful, especially after feast days, and vellum was not; and then he had been taught how to scrape and cure sheepskin to make his own parchment. He enjoyed all this, and found it easy, until one day his father, a priest, came upon him and a young male relative, a chorister, concentrating on a long scroll held down between them: crude archangels, demons, horned women; spells in angular letters, all red. How dare you! His father's hand had twisted his ear until it burned. How dare you corrupt your learning, your soul, with – this, this dragging of Satan out from where he belongs! I forbid you to pick up a pen and write, ever again. May curses rain down upon you if you even think of tracing anything other than your name!

All the students had to learn most things by heart, but after that Tsega had to commit everything to memory: the divine offices and the book of hours, the antiphonaries, all of David's psalms. When he graduated to the school of qiné, church poetry, he pulled his head through a rough sheepskin cape, picked up his leather book case, and left for a nearby village, where he had heard a respected teacher was working. A handful of others had done the same, walking in through the valleys and the mountain passes, choosing mastery of poetry in Ge'ez, the church language, over homes that they often never saw again. For five years the sun rose to find them gathered around their teacher, listening to him describe stanza forms, explain particularly pleasing metaphors, recite useful examples. They memorised model qiné and with his help peeled back their punning layers, looking for the gold hidden within the wax mould, the meaning nestling at the centre like the dark hard core of an olive tree. The church told them they were training their minds and souls, opening themselves up to appre-hensions of divinity, but Tsega was learning worldlier things too: how to smuggle deniable meanings into seemingly innocuous conversation; how, because qiné carried with it so much prestige, it might be a way for a village boy disinclined to soldiery to chase social advancement.

During the day the students scattered across the countryside, composing their own poetry and begging, as the church provided no food. Tsega hated this aspect of his calling. He was proud, afraid of dogs, and quickly resorted to tall heart-tugging tales. In the late afternoon the students returned to their teacher, who listened to their verses, then easily, deflatingly, disassembled them. Near the end of the five years Memhir Hiruy, famed throughout the country for his skill with qiné, visited the Gojjam school to teach. The students vied amongst each other to impress the master, who after a couple of weeks singled Tsega out for praise. Would he like to come to Gondar to continue his studies? Of course he would. And, ignoring the protestations of his mother, he went.

Not long after they arrived at Ba'ata, the priests asked Memhir Hiruy to perform a qiné. 'Ask him,' replied the scholar, pointing to his new acolyte. They were insulted. Recent experience had only confirmed their deep suspicion of outsiders. And who was this anyway? A youth from the sticks – from Gojjam, no less, where everyone knew the evil eye flourished. Why was he here, assessing them with his noncommittal gaze, threatening them with his very presence? Why, he hadn't even finished his studies. But after a glance at Memhir Hiruy for reassurance, Tsega stepped forward to puncture their scorn.

When Memhir Hiruy left for the new capital, Addis Ababa, Tsega stayed, learning the recondite church dances and committing to memory all the books of the Bible, the Old and New Testaments, their interpretations and commentaries, and the books of the Fit'ha Negest, the law of kings handed down from Byzantium and from medieval Egypt. He became a teacher himself, and, ambitious in the way of people who know they have only themselves to depend upon, quietly but steadily made connections, travelling across the city after church services to read to the families of increasingly important personages; for Tirunesh's husband, among them, and for her oblivious niece.

Less than a year after his wedding, Tsega deposited his young bride at her grandmother's house and was on his way south. The mules were watchful on the long trek down into the Lake Tana basin. Their riders, too, looked around, into shaded copses, up at the lips of ravines, and once they had arrived at the Blue Nile gorge and were picking their way down its steep sides, into anything that even suggested it might be a cave. Everyone knew, from childhood stories, from scarred survivors, that this fertile country ran with bandits who regularly stripped mule-trains of their valuables then pushed off in low reed boats, poling them through the mud-coloured lake to islands and promontories, or disappeared into caves. They were all grateful when they scrambled up onto the wide cool highland plateau and then down, through aromatic juniper and newly imported eucalyptus, down into Addis Ababa.

Once there her husband took his time, acclimatising, visiting Memhir Hiruy, attending services, listening to gossip about the empress, and especially about her subtle regent Ras Tafari, who had recently returned from an extended tour of Egypt, Jerusalem, and the European capitals (where, among other things, he had wisely declined to sign a treaty that would have allowed Italy to build roads, rails, and a port into Ethiopia). The regent also managed – daily, it sometimes seemed – to announce the institution of new-fangled things: a modern school, a printing press, a newspaper. Every decision of any importance passed through his hands, which was a useful thing to know, but for the moment was not what most interested the new-minted priest. Tsega presented himself at the head offices of the church, and was given the care of a country parish called Gonderoch Mariam. He travelled the city to read aloud in the households of great men, among them Ras Kassa, the regent's pious cousin. And he joined the throngs that arrived daily at the palace, looking for an audience with the empress.

Menelik's daughter Zewditu, crowned after the brief reign and ruthless deposition of Menelik's grandson Iyasu, was phlegmatic, conservative, uncomfortable in the presence of men and overwhelmingly pious, and had only once bowed to pressure to preside at the courts that operated in her name; she had hated the work, and never returned. She preferred, wrote her chronicler, kindly, to confine herself to 'spreading spiritual wisdom by fasting, prayer, prostrations, and by almsgiving'. She read the histories of the female saints, 'and a spiritual envy [to be like them] was stamped on her heart'. Each day she rose early and prayed into the afternoon, eating nothing, outdoing many of her monks and nuns.

Zewditu's self-denial was matched by generosity – socially required, an expression of pride and status as well as charity, but prodigal nonetheless. Hardly a month went by without a banquet given to soldiers, to clergy, to the nobility or the impoverished laity, and late one November, after Tsega had been in the capital for nearly two years, he was invited to one. The floor of the great hall was laid with hundreds of carpets, some of wool, others of silk. Guests filled the vast space, seated according to their rank: the empress, her regent and senior princes of the blood on a raised platform at one end, surrounded by curtains, then, when they had eaten, the curtains drawn back so they could look down at long low tables filled with lesser notables, ranks of clergy in high white turbans and glowing white shemmas, straight-backed soldiers. Crosses blinked in the lamplight, phalanxes of servants and slaves brought horns of mead and baskets piled high with injera. The smell of dark red chicken stew; of zign, beef in ginger and cardamom and bishop's weed; the sight of entire sides of fresh-slaughtered oxen carried on poles balanced on the shoulders of slaves so anyone could take a knife and help themselves, made him ravenous, but he ate nothing at all.

Eventually one of the higher-ranking servants enquired why. 'Because this is the day the Ark of the Covenant, captured by the Philistines, was returned by God,' he replied. 'I am fasting in celebration of Zion.' The servant, knowing this was the kind of thing that interested Empress Zewditu, told her there was a priest in her hall observing the fast of Zion, and who therefore could not partake of the abundance of meat on offer.

She turned out to be observing it too. She had had fasting food cooked for her, pulses and vegetables rather than meat, and she sent the young priest a portion of it. When he had eaten he stood and addressed to her a qiné of praise he had composed in preparation for just such an eventuality, a poem playing upon the biblical echoes of her baptismal name and lauding her holy magnanimity. She inclined her head in thanks. 'And what can I do for you?'

This was why he had come to the capital, the moment he had been working toward for years. 'I would like to lead Ba'ata Mariam church in Gondar,' he replied. 'And I would like to rebuild it to the glory of God.'

'Of course,' she replied. And she ordered the provision of all the accoutrements the new aleqa would require: a cape worked over in gilt, a tunic with a wide coloured band embroidered at the hem, a sash; bags of Maria Theresa silver.

Half of Gondar, it seemed, came out to meet him, ululating praise of their new chief priest. Aleqa Tsega accepted the celebrations calmly, savouring his sudden leap above those who had denied him welcome, noting the practised tributes from his fellow clergy, the underlying silences, the curdled smiles.

At her aunt's house a feast was waiting. There too he looked about him – at the told-you-so pleasure of Tirunesh, the reluctant approval of Mekonnen. At the narrow-hipped girl in black who hung back, shaven head held low.

TIQIMT
THE SECOND MONTH

Sunny growing and ripening season. Honey removed from hives. The first barley threshed and winnowed for roasting and beer-making. Children play outdoor games; girls dance and greet storks arriving 'from Jerusalem'.

She no longer hid as she had before he left, running to the store-room, burrowing in among the wheat and split peas, the dusty green-smelling, crackling hops. Breathing shallow so no one could hear, crouching down behind a high basket or clay water-pot, willing herself invisible. If he found her he had chastised her, playfully. What, hiding again?

Now, nearly four years older, she did not hide, but still she retreated to the back rooms, to sit on a low stool among the comforting grains and watch the days crawl across the floor. She would have spent the nights there, if she could. She did not look up at him, or speak to him; even if she had been expected to, she could not.

When he was out the servants took charge, bossing her about the house like the child she still was, letting her help, yet refusing to play games with her because, being married, she was no longer a child. So she played alone, making a head from bunched-up cloth, a body from a dress, and rocking the form to sleep. Sometimes she heard children's voices beyond the walls, whispering, calling – 'coo-coo-loo!' – and swallowed the voice that itched to answer.

Other times she sat at the window, craning for glimpses of life. Morose donkeys clopped by, or women doubled over beneath wide loads of firewood. Slaves with high packs balanced on their heads, nuns in yellow caps, children running errands. Once she saw a great lord riding in the direction of Ba'ata. His mule clashed and jingled with embellishments and the sun lit the dull barrel of a rifle. Retainers scurried to clear the way.

How handsome he was. 'Who's that?' she asked the slave girl, who shook her head, and ran to see.

'Ras Gugsa,' she reported. Ras Gugsa. Their governor! From her father she knew she was distantly related to him, and even here, shut up among the servants, she had heard the rhymes and the gossip. He was pious, as required, a poet and a fair administrator, but somewhat hidebound, too, and a melancholy and determined drinker. He had been married to the empress, who, it was said, still loved him, but they had been forced to separate when she was crowned; it was no secret he blamed his loss of power on the regent. Just over a year later he would be tricked into battle against Ras Tafari and die on the fields of Anchem, his soldiers having scattered in fear of the regent's most recent toy, the aeroplane. But for now he carried all the sheen of high office.

When they were first married her husband had hired a blind abba to lead her, singing, through the alphabet, the set texts of early church school, the psalms her mother had hummed to her. She found the abba kind, loving, but soon he was reporting that she was too impressionable, too prone to tears. And too quick to learn, too. 'If you correct her or do her wrong,' he said to her husband, 'she will quote David at you. She will cry to God will listen to her. Do not teach her to read.' So the lessons stopped, and she sat out her hours spinning thread from tight bolls of cotton, twisting coloured yarn around narrow bundles of straw to make serving baskets, or picking crumbs of dark earth out of quintal after quintal of wheat kernels, lentils, teff.

Sometimes, still, in a sudden access of spirit, she would run to the neighbours', climb up into their peach tree, fill her skirts with the biggest fruits she could find, and slink back to enjoy them. Once she left the main door open by accident. A sheep had just been slaughtered and a dog crept in and got hold of one of the back legs. Heart thumping so hard it seemed it might deafen her, she managed to startle the animal into dropping it.

Other times she acted willingly enough in the play that had been written for her. Not that she necessarily knew the words, or her exits and her entrances. So at harvest time, after the peasants had delivered their tithes – two-thirds of the barley and wheat from the Jews who farmed at Gonderoch Mariam, the smaller church her husband administered; chickpeas and chilli peppers, peas and broad beans and teff from Ba'ata's lands in Dembiya and Bisnit – she handed skiffs of wheat and barley, balls of butter, strips of beef jerky and cobs of corn out to anyone who looked as if they needed it. There was so much she felt it wouldn't be missed.

Or guests dropped by. 'Where's your father?' Sometimes she could not help but laugh. 'Oh, *you're* mistress here!' By the time she turned twelve she was becoming accustomed to being called woizero. Lady. Enjoying it, even.

As such she was not expected to grind grain or collect water, but she was expected to be able to cook, to provide handsomely for the priests, the merchants, the visiting dignitaries her husband brought to the house almost daily. Sometimes, knowing her instruction had been interrupted, he helped her, tasting, suggesting, demonstrating, assuming she knew this had to be a secret held between them lest it diminish his station. Until one lunchtime he criticised her: the fish in the wat had been overcooked and was breaking up, there was not enough sauce. Child, he said, this is a bit dry. But we made it together! she protested, before she could think.

The next time she was brought fish, five fresh silvery creatures from the Angereb river, she was extra careful, stripping them, washing them, removing every bone, rubbing the pieces with spice. The resulting wat was succulent, perfect, and when that day's guests arrived, and took too long at their conversational preliminaries, Tsega was impatient. 'Never mind, sit down and eat.' There was no feigning their enjoyment of what had been put before them. After they left he took her small hands and kissed them, over and over, until she thought he might swallow them.

He took a keen interest in her deportment. It wasn't enough to wash her hands before and after meals; she had to scrub her arms up to the elbows. When her official mourning for her mother and brother was ended she had begun to grow her hair again and braid it back from her forehead. Other girls put silver rings in the plaits, but he would not allow it. Soon he found even the shining braids too much, and told her to hide them under a scarf.

She understood her new state meant she was to stay at home, but initially she did not understand how absolutely he meant it. She had always previously been allowed to run over to a neighbour's to borrow pots, or muslin to strain butter, or a few shallots, and she still did so. One afternoon when she returned, however, he was waiting for her.

Come here. Her stomach seemed suddenly to have slid to somewhere around her feet. Come here, I said. He raised a stick, and he did not stint. At first she was so shocked she could not cry, but then the sobs arrived, deep and gusty so she could hardly breathe.

But with him it was as if a tempest had passed. Anxiously he stroked her head and picked at her shawl, straightening it, smoothing it over her shoulders. My heart, don't cry. Don't cry. Here. Here's some money. Pressing silver thalers into limp hands. Get the servants to buy you something nice. Not jewellery, you know I don't like jewellery, but something nice.

Sometimes he worried whether she ate enough. Lijé, he'd say. My child. My child is hungry. And at night especially, when there were no strangers about, he would draw her close and feed her from his side of the mesob. The portions were too big, so she would intercept his hand and break them up into smaller pieces, eating what she could, then closing her mouth tight.

Not infrequently he would arrive home to find her in a corner, weeping. Child, he would say gently, why are you crying? Who has harmed you? And at last the answer would come. My mother. My mother is dead.

Ayzosh, ayzosh, he would murmur, drawing her to him. He dipped a hand into a wooden vessel that had held butter from Asmara. When he drew it out it glistened with the remains of the butter, and with it he would wipe away her tears and gently soften her taut and salty face. Ayzosh. I will be like a mother to you.

After one of these moments he seemed to be concentrating on her longer than usual, drawing dark fingers down her neck. They stopped at the centre, traced a spot low on her throat. Are you growing a goitre? he said, almost to himself. She had little idea what a goitre was, so as usual she said nothing, and soon forgot he had asked the question at all.

But some days later a servant came to her to say, that lady the master asked for, she has come. What lady? But she greeted the woman, and watched as the woman set about heating oil-seeds over a low fire, stirring them until they smoked and burned.

Watched as she scraped the soot off the sides of the gas lamps and added that to the black residue. A bit of kohl, too, so the mixture glistered and plopped on the heat.

The woman set it aside to cool, then walked over and took her by the hand. 'Now, sit still.' She took up a narrow stick, dipped it into the cooling mess, and began to draw a line around Yetemegnu's neck, parallel to her collarbone. As suddenly as she understood she was on her feet. But the servants held her down as the woman drew another line, and then another and another, and at the ends of the longest, just under her ears, risen suns.

'Araqi?' Alcohol would numb her, but she could not assent to any part of this. She shook her head, a sharp snap of refusal. 'Don't move!' She closed her eyes.

When the needle punctured her skin, she exploded, biting and scratching and writhing. But they held her tight, across her body, by her head, so nothing could stir – only her tears, streaming down her face. And her mouth, screaming. What had she done to God to deserve this?

After the scabs had hardened and fallen off, after she had spent two months delirious and burning with infection, she looked and saw she was imprinted with the tracks of her own tears.

In her first pregnancy she slept all the time. They fed her and she accepted it, and they fed her more. And when finally they wrapped her up tight and sent her on slow mule-back to her father's home in the countryside she felt only relief, to be away from the big house, even though her husband had already left some months before, taking the road south to Addis Ababa as soon as he had heard word the empress was dead.

One morning she felt a trickle down her thighs.

The women gathered, aunts, grandmother, neighbours. None seemed especially concerned, though they did become quite busy. Some raised a dividing curtain across the main room, others began to roast and pound and boil coffee. Charcoal was brought in on a small stand and blown into redness. Incense curled into the rafters. Watchful laughter, and chat.

Yetemegnu, now fourteen, had been told birth would feel something like it did when she went to the grove behind the house at dawn and squatted to relieve herself, so she had thought of it as that painless and that quick. All she could think, as the contractions tightened and tightened their grip, was no, this didn't feel like that at all. She gasped, and the women's voices rose to meet her, to share her pain, to distribute it between them.

Mariam Mariam Mariam Mariam Mariam, dirèshilin.
O Lady Lady Lady Lady Lady, come to our aid.

Outside the sun shone on fields vivid with young crops, yellow oil-seed, blue flax, the nodding, dewy greens of new barley, broad beans, wheat. The air, washed bright after months of storms, picked out every tree on the hill that rose behind the house, every silver leaf in the eucalyptus brakes.

But inside it was dark with people and smoke and low talk. In the doorway, behind the curtain, a deacon read from the homilies of Ruphael. Listen, the women said. Listen, because Ruphael opens our wombs.

She sat at the centre, on a low stool. One woman stood behind her, strong arms clasped across her narrow chest. Another sat at her feet. They told stories, asked her questions, tried to divert her, but she sank further and further into herself, dreading the next visitation.

Mariam Mariam Mariam Mariam Mariam, dirèshilin.
O Lady Lady Lady Lady Lady, come to our aid.

They told her she was not meant to cry out, but she couldn't help it. Oh Mary, mother of God, relieve me! Ayzosh, they said. Ayzosh.

She made wild and breathless promises, about the prayers, the fasting she would undergo if only this could stop. One of the women laughed. Oh, you'll forget. And you'll be having another soon enough.

The deacon read on, a baseline whose timbre changed only when he shifted in his seat, or coughed.

Mariam Mariam Mariam Mariam Mariam, dirèshilin.
O Lady Lady Lady Lady Lady, come to our aid.

An endless night, and another day. The women had put down a mattress so she could curl up on her side, but she was now back on the stool. They took turns holding her and drank cup after cup of coffee, but no one ate.

The deacon read on.

Another night, and another day. St Mikael's Day, bathed in birdsong and sunlight.

Eventually the deacon put Ruphael aside and took another battered book out of its hide case.

'NOW THERE WAS A CERTAIN CITY WHEREIN A CHURCH HAD BEEN BUILT, AND THE CHURCH WAS BUILT IN THE NAME OF THE ARCHANGEL MIKAEL, AND EACH YEAR, ON THE TWELFTH DAY OF THE MONTH OF HIDAR, WHICH IS THE DAY OF THE ARCHANGEL MIKAEL, GREAT NUMBERS OF THE INHABITANTS OF THE CITY DID NOT FAIL TO VISIT HIS CHURCH – MAY HIS

INTERCESSION AND HIS SUPPLICATION KEEP OUR KING DAVID
FROM THE EVIL ENEMY!'

The baby crowned. She was beyond pain, beyond comprehension.
She felt her spirit departing, the world about her fading. The
women's voices rose.

Mariam Mariam Mariam Mariam Mariam, dirèshilin.
O Lady Lady Lady Lady Lady, come to our aid.

'AND BEHOLD IT CAME TO PASS ONE DAY WHEN THE PEOPLE
WERE JOURNEYING ALONG THE ROAD TO COME TO THAT
CHURCH, THAT A MIGHTY ROARING RUSH OF WATERS CAME
FROM THE SEA, AND IT BURST UPON THE PEOPLE AND ALARMED
AND TERRIFIED THEM EXCEEDINGLY, AND DROVE THEM OUT OF
THEIR SENSES; AND THE WATER SURROUNDED THAT PLACE AND
ROSE TO THE HEIGHT OF ABOUT TWO MEASURES, AND THE
PEOPLE WERE WELLNIGH DROWNED.'

And with a great tearing and an onrush of fluid, the baby arrived,
pouring into the arms of the midwife at her feet. She fell back and
they caught her and laid her down to rest.

When the placenta had safely followed they took a pat of butter
and rubbed the newborn's head. They pinched its nose, smoothing
it, pulling at it, so it would grow long and narrow. And their ililta
rose, loud and joyous.

Nine times. A messenger was dispatched to Addis Ababa, to tell
Aleqa Tsega his first child was a girl, and that she was named
Alemitu.

In her father's village the feasting began. Araqi and tella flowed
like water. Injera arrived by the mesob. Five sheep were slaughtered, and then someone brought four more. All who attended the

birth and fasted for the duration ate their fill. But the new mother took nothing at all. They gave her honey, and buttered oatmeal gruel, but she would not touch it. She lay curled on the floor, hugging herself, trying to sleep.

It was only when, decades later, she herself beat through the streets of Addis Ababa in supplication that she began to get an inkling of where her husband had been when Alemitu was born: under a colder sky, walking raw avenues that criss-crossed the bottom of a green bowl surrounded by mountains. The city had been capital for only forty years, and they were still building it. Donkeys trotted through the main streets as they would in any town in the country, carrying firewood, dried discs of cow-dung, baskets for market, but here they also hauled scaffolding, sheets of corrugated iron, drew telegraph poles rattle-bouncing through the dust. Everywhere a new structure was going up, a road being metalled, cable installed, or a ragged work gang being whipped into a semblance of efficiency. Rumours eddied through the thoroughfares: that Tafari was overseeing every detail of the preparations himself; that every day he appeared, a small long-nosed black-caped figure, at a hotel, or the market, or the telegraph office or on the wide sweep of road leading from the railway station, checking on the quality of asphalt or of uniform or of welcome; on the impression his city would make on the foreign guests who had already begun to arrive.

As the coronation neared, the donkeys were joined by horses and mules, trotting footsore in from all across the empire. They carried warriors and their generals; lords of the north, the south, the east and the west, princes who owned more land than the regent, who commanded armies as big as or bigger than his.

Princes and priests, priests and more priests, from Aksum and Debrè Libanos and Meqellé, from Sidamo and Harar and Debrè Marqos, or, like her husband, from Gondar, priests in white turbans, black capes, long self-regarding beards; monks and nuns in turmeric yellow, gold-red amber necklaces hanging heavy on breastbones. Prayer sticks and crosses everywhere, crosses of gold and silver, of brass and of humble wood. And Tsega moved proud amongst them, listening to them jostle over the niceties of hierarchy and opportunity afforded by Tafari's achievement the previous year, of finally beginning to bring independence to their sixteen-hundred-year-old church.

Over sixteen hundred years since a ship on the Red Sea coast was attacked, leaving two Christian boys from the Phoenician city of Tyre to find their way to the Aksumite court where they became companions of Ezana, the emperor's son. Sixteen hundred years since one of the brothers, Frumentius, now grown, travelled to Alexandria to tell the leadership of the Egyptian church that Christian merchants from across the Roman empire congregated in Aksum and needed a bishop. Frumentius himself seemed best qualified for the job so they consecrated him and dispatched him back to Aksum, where he converted his childhood friend (who, now emperor, well understood the usefulness of such ties with Rome, his chief market for ivory, tortoiseshell, gold, rhinoceros horn). The Ethiopian church had been led, in name at least, by an Egyptian monk ever since, but when Empress Zewditu promoted Ras Tafari to negus, or king, of Gondar, he had moved at once to free it. Assiduous diplomacy produced, within a year, a promising first step – division of the empire into dioceses led by five Ethiopian bishops, among them Abunè Abraham of Gojjam and Begemdir.

*

Eventually, seven days before the coronation, the milling priests resolved into an expectant phalanx. Serried rows of eyes, her husband's among them, watched as an honour guard marched through the gates of St George's Cathedral and up the steps. Rows of eyes watched the reverent handing over of robes, of a sceptre and an orb, of gold rings and spears, of a sword sheathed in gold and diamonds, of a gold-encrusted Bible and two heavy gold crowns. Then, for seven days and seven nights, seven times seven priests sang David's psalms and the Book of the Praise of Mary, cycling again and again through all their strophes of light.

The words, of course, were familiar to all the men, who had each spent decades studying their order and their intent; so familiar they were all too often muttered and gabbled, boredom taking precedence, minds drifting to other things; but when, in the dark hours after midnight on the seventh day, Negus Tafari and his wife arrived at the cathedral to begin their vigil, it was as though the phrases had been renewed, charged with present meaning. Also that of the past: this coronation drew much of its power from the Kibrè Negest, the Book of the Glory of Kings, in which, as all here knew, it was told how Solomon of Israel was visited in Jerusalem by Makda, queen of Sheba. Solomon, who would eventually number among his consorts seven hundred wives and three hundred concubines, was attracted to Makda, but she was a virgin and disinclined to yield, so he tricked her, feeding her spiced food and ensuring the only water in the palace was by his bed.

Solomon and Makda conceived a child, whom she called Menelik. Thus, claimed the chronicles, did the pearl God tucked into Adam's body at the beginning of the world, the pearl which had been passed down through the children of Abraham to David and to Solomon (and would eventually arrive in Mary, and in her son), pass into Menelik also. When Menelik was grown, he travelled to see his father in Jerusalem. Solomon anointed him king of Ethiopia despite a troubling dream he had had, of a bright sun blazing over Israel then departing south. When Menelik I at last left for home he did so in the dead of night, taking the Ark of the Covenant with him.

As dawn slipped through the high windows and traced bright lines across the carpeted floor, cutting through thick eddies of incense, dispersing the blurred yellow light of hundreds of bees-wax tapers, the chanting and the prayer intensified. The deacons danced before their king, before the holy of holies and its sancti-fied replica of the Ark, as if for the first time; dancing as David danced in the temple of Jerusalem – sistra clashing, drums beat-ing, bare feet stepping, serious and joyful.

Outside the cathedral two open-sided coronet-topped tents had been erected within a far larger tent that held the most important members of the congregation. This tent was filled with alien rega-lia, for apart from the great rases and court officials, few Ethiopians had been invited. When the priests accompanied Negus Tafari out of the church, they stared amazed into this spectacle. And the spectacle stared back. Politely it looked, and less politely it assessed: strengths, weaknesses, allegiances. Behind puffed-out chests and polished medals, under topis and busbies, calculations in English, Italian, French, German; calculations of land, export capabilities, porosity of borders, military prowess. Many of their Ethiopian hosts knew this, and if they didn't know, guessed, for how did that proverb go? Oh yes – foreigners enter like thread into a needle then branch out like a sycamore fig. Why, only six years ago Ras Tafari, touring Europe, had had to use Ethiopia's new membership of the League of Nations to shame Britain and

Italy into dropping a plan to divide influence over his country (they drew up another a year later). Italy, surely, should have known better: had it really forgotten Adwa, where, only thirty-three years ago, Menelik II and his empress Taitu had routed the Italian army? No, no one had forgotten that; some sitting here had even fought in that war. The Italians had had to content themselves with retreating to their colony in Eritrea, and with fantasies of revenge.

All the more important, then, that the foreigners should see this show of pomp and power. And that they should be well acquainted with this new leader, who had so steadily extinguished all internal opposition: in war, as had happened with Ras Gugsa; around the council table; by sheer attrition: Empress Zewditu, for all her supposedly final word, and for all her manifest reluctance to promote her busy regent to negus, had eventually found she had no other choice, for he worked as tirelessly and invisibly and patiently as the weather. And then she had died, suddenly.

Now Negus Tafari bowed low, touched his forehead to the stone of the cathedral, kissed it, bowed, kissed, stood. 'I will lift up mine eyes unto the hills, from whence cometh my help.' A clear, controlled voice, with a hint of a rasp. 'My help cometh from the Lord, which made heaven and earth.' Archbishop Kyrillos moved to the table to begin the coronation proper. 'Now, according to God's will and goodness I am going to crown him, and anoint him king of kings, so he will work with all his body and soul to spread religion and increase education, and Ethiopia will rise in wisdom and in knowledge, and her flag will be laid down from border to border. And we charge that you will be ruled by him and help him in this good work.' When the reply came, in a rumble from all around, from the chairs, from under the trees, from among the graves, 'May God help our emperor do as you say. Amen,' he placed his hands on the Bible and turned to ·

the king. 'Will you, in your authority and power, and in all your works, watch over the people of Ethiopia with patience and compassion, and keep their wellbeing in mind always, according to the law?' 'These words shall lead to good works, so, insofar as I am able, yes, I will.'

And so the service, in which each accoutrement – sceptre, orb, spear, ring, crown – was blessed by the archbishop and by a scholar of the north, of the south, of the east and of the west, a service interspersed by the voices of ten deacons handing their chants back and forth, back and forth, unspooled with the solemnity and intimacy of a wedding. (Though no weddings would have been accompanied, as midday approached, by a vast roaring in the sky as Tafari's beloved aeroplanes swooped and looped in fealty.) As in a teklil wedding the vows were followed by a mass; as in a wedding the ceremony included communion, for which Tafari and his wife were required to return to the sanctuary and replace their rich robes with something simpler. As in a wedding the service was brought to a close by qiné after rich dense qiné, one of which was delivered by Aleqa Tsega. And then Emperor Hailè Selassie I, Conquering Lion of the Tribe of Judah, King of Kings of Ethiopia, elect of God, climbed into a horse-drawn carriage and was driven away.

When at last Aleqa Tsega returned to Gondar his coming was heralded by a warning: you are the wife of a great man now. The emperor has tied a circlet of gold about his head. When most of the foreigners had departed and the daily banquets and firework displays were tapering off Aleqa Tsega had answered a summons to the palace, where he joined whispering huddles of clerics and lords, all wondering why they were there. Promotion, it transpired: Emperor Hailè Selassie, who knew well how favour, generously bestowed, tightened the reins of loyalty and obligation (especially in those of humble birth), was parcelling out authority over sections of his realm. Aleqa Tsega returned a liqè-kahinat, chief of the learned, of all of Semien and Begemdir. But at the time she understood only that she would have been happier if he had not come back at all.

BOOK II
1931–1941

HIDAR
THE THIRD MONTH

*Ripening of later cereal grains, sunny, occasional
mists. Harvest, especially of early teff. Boys take
cattle from sun-scorched uplands to green valleys.
Flirting season.*

*Mariam Mariam Mariam Mariam Mariam, dirèshilin.
O Lady Lady Lady Lady Lady, come to our aid.*

Her next labour went quickly, and twelve ililta rang out before
dawn. A boy.

She basked in congratulations. Her husband basked in
congratulations.

On the third day a family friend came to visit, ululating. This
friend knew how the first birth had gone; that Yetemegnu had not
been able to sit for months. That they had kept trying to feed her
barley gruel sweetened with honey, and when that failed honey
with water, but she didn't like honey, and again and again she had
closed her lips tight and turned away.

This time they brought her wheat porridge full of fortifying
butter, and this she had accepted.

She smiled at their friend. The boy has not yet been washed –
would you do the honours? His skin is darker on one side than the
other, don't be surprised.

'Of course.' And their friend had fetched Lux and water and
rearranged her shawl, freeing her arms to work.

'Oh!'

What is it?

'She's a girl, not a boy!'

No. No – can I see?

The child's umbilicus had been cut long, and an attendant,
glancing cursorily, taken at her mistaken word.

Oh no! The father was so pleased to have a boy! *I* was so pleased
to have a boy!

And then, desperate, How do I tell him?

Their friend considered her. 'Take your time, and do it carefully.' She smiled. 'Wait until he's had his lunch, at least.'

Preparations for the christening party were nearly
complete when, just under three months later, the girl died, under
a light shawl Yetemegnu had placed over her to protect her as she
slept.

They had left the house near Ba'ata because there was no water. No
pipes serving it, no springs save the holy springs, no well. Only the
Qeha river, close enough as the kite flies but down and then,
carrying the big-bellied madigas on their backs, back up such
steep paths that the slaves she sent were exhausted for the rest of
the day. Sometimes she took pity on them and went to the water-
sellers instead. Then they heard that a man who owned a plot of
land on the edge of the Saturday market had committed a murder
and fled. They were all pleased when her husband paid a messenger

twenty pieces of raw silver to track the murderer down and make him an offer. Sixty silver thalers for the land, then; and a hundred to sink a well.

Her husband built a house of hewn stone, held together with trampled mud and straw, with stairs at each end and in front a round hall in which to receive visitors. A little building off to the side for grinding grain, and on the ground floor of the main structure rooms to keep it in; rooms for fermenting beer, for storing mead, and above them a private living room and their bedroom, with low wooden chests and their clothes hanging on hooks on the wall. The windows on this floor faced sunset and sunrise, and when she was alone she could stare out the west window, over the roofs and into the valley, where in the dry season the Qeha crawled sluggish with algae, and in the wet season rushed wild, a muddy and perilous torrent. If she turned a little, toward the north, she could see a darker circle of high old trees, which always meant, here is consecrated ground.

If her second daughter really had been a boy, he would already have been baptised, and named, at forty days; girls, however, were baptised at eighty, and so the tiny body had had to be laid in unmarked earth just outside Ba'ata's walls. She had not been allowed to attend the burial, mothers in her position rarely were, and consequently she had been denied a formal farewell. Now, standing at the window or bending over the fire or walking through the wide rooms of her new dominion, she often found herself in tears. Sometimes she knew she wept for her unnamed daughter, lying in the cold outside the church, sometimes for her mother, sometimes for herself. Sometimes she was not sure what she wept for, only that the tears leaked silent down her face and would not stop.

Of course she also knew how often deaths like this happened – why else were mothers not allowed to leave the house until the eighty days were up, but to keep a baby safe from illness, from strangers, from the evil eye? Except she felt that it was she, sitting at home, who had caused the danger, and she who now checked and rechecked every blanket, every shawl placed near Alemitu, waking in a panic in the middle of the night to check again. Why else did fathers so often keep their distance through the first years, afraid of attachment to a child who might not survive? Well, there was no problem with that – she kept her distance from him, speaking only when spoken to, and often not then.

But at fifteen she was still so young, and Alemitu – raised on rich butter from Qimant country, and still feeding from breasts now heavy with milk for the absent child – Alemitu was growing fast. They played together, running through the wide yard, and she would laugh at the attempts at language, laugh at the serious, halting explorations, slap hands away from danger, wonder at each small important thing learned, existing only in the bright present of both their childhoods.

And there was now so much work. Every morning she woke as the sun began to crest the mountains. Quiet, in near-dark, she would slip out from under her blankets and shiver over to the madiga, dipping a gourd into cold clear well water and splashing her face awake. From the silhouettes of the other houses would come the scrunch of stone against stone as the first grain was ground, and the smell of sleeping fires blown back into flame. Roosters crowed into thin mountain air. The occasional dog barked. Figures wrapped tight against the chill hurried back from the woods.

Her husband would be up by then too, sometimes taking breakfast, more often not, pulling on jodhpurs, a long shirt with tight embroidered wrists, full black cape, turban. Other priests took their time, winding the lengths of white cotton just so, but he was impatient with it, and she, who was so particular and neat about her person, would glance at the mess on his head in silent irritation.

If she ate breakfast she ate it alone. Light began to fill the big house, and the sound of chopping; the smell of onions frying, and Alemitu running, or crying. Sometimes there would be the guttural bleat of a sheep, suddenly cut off, or the flapping screech of a chicken losing its head, but more often it was pulses and spices and vegetables that required cleaning, pounding, mixing, bubbling for hours over the fire. A pause at nine, for fresh-roasted coffee with the servants, a basket of delicate sorghum popcorn, a knobble of incense dropped onto red coals, and then it was back to the dark kitchen hut, to lean over the flat clay pan, pouring thin sourdough in quick, even spirals so it made the thinnest, whitest possible injera.

Just before he came home at one she would take the ilbet from the side, where it had cooled, and beat it into a savoury paste. She would heat the sauces she had made that morning, roll fresh injera – always far more than she needed for the two of them, because she never knew who he was going to bring home. She never sat with them, of course, but as she came back and forth from the kitchen, serving them, she caught snatches of talk. Church politics, the whereabouts and doings of the new emperor. Those Italians at the new consulate up the road: employees were given better housing and their own school, and many townsfolk were envious. Now someone had attacked the foreigners' telegraph station, because, they said, it was the work of the devil.

Occasionally she raised her eyes from the ground or the dish she was serving to glance at her husband. She saw that he was a listener, and a watcher, a man who knew the power of silence and of a quiet, steady voice. She saw his thoughtful courtesy, in these days when he was working assiduously to consolidate his position; saw his veiled pride; knew, even if he did not show it, his foreigner's anxiety, the insecurity of recent arrival (though he had now been in Gondar for over two decades), his need to prove himself better, worthier, accepted.

He was not always home. He had obligations at Gonderoch Mariam, where he led the church as well as owning land; in the capital, where it was useful to keep his face fresh in the right minds; for one too-short month in Gojjam, where, much later, when she actually cared enough to notice and ask about such things, she learned he had pitched a tent outside his childhood church to teach poetry and church administration.

Her own limits were established now – these cool rooms, the expanse of fenced ground around them, the narrow gateway through which she could see snatches of activity in the market – and she no longer dared to venture beyond them. She did not invite anyone round – and anyway, who would she ask? He could not forget how her father had so looked down on him, and now they had their own house he had made it clear her family were not welcome. Neither were girls her own age, being potential agents of temptation.

But local matrons sometimes bowed through the doorway, nuns, a kind hunchbacked priest, and she sat before them, heels tucked under her, pouring coffee, occasionally suckling the child, silent but for murmured rote responses, while in her mind questions nudged toward clarity. There was tradition, she knew that – how many times had she heard a chuckling elder say it, that women and donkeys need the stick? Then there was the fact that, as her husband had pointed out more than once, you were an unformed child entrusted to my care, and all children must take correction. There was their specific circumstance, which over and over he explained. You cannot be gadding about and blackening my name. A priest must be blameless, and so must his family, they cannot be accused of luxury or lechery or stubbornness, they

must be seen to be obedient. You cannot damage m̲
And: our vow was for life, made in the church; you risk your
And: aqwatiré yizishalehu. I will care for you, look out for you; I
will gather you up and hold you close.

One afternoon when she was about nineteen, two young men,
strangers, came to see her husband on business. She was in the
back yard when they left and one of them darted over to grab her
hands in his. 'Come, come with us.' Startled, giggling, she snatched
them away. Where? Why? Unable to explain to someone so
uncomprehending, he dropped her hands and took hurried leave.
Did you see, master? she said later. How good-looking they were?
She meant it innocently, would have said the same, she told
herself, if they were women and beautiful; she was admiring God's
work, what else could she possibly mean?

Why did you not go with them, then? He smiled, but at once
she saw herself through his narrowed eyes, and through theirs.
Long lashes, decent teeth, a slender, mobile, fertile body. A high
forehead and long nose. Skin alight with youth.

The next time he punished her she ran away. Ran panting to
her grandmother's house, was received with hugs, with good food,
with comfort – Ayzosh, enaté, ayzosh, my heart. Ayzosh. And was,
a few days later, returned to her husband. You are friends again
now, aren't you? Good.

Again she slept a great deal. She prayed. And she dreamed.

One night she dreamed that the bishop Abunè Abraham came
to her, slipping into the bedroom at the top of the house. She
watched him, unable to stir or speak. A shiny black hat, a nice
face, she thought, at least from what she could see between hat
and beard, but she was afraid to look too close. There was a huge

metal-bound chest in the room, a present from her grandmother, who had bought it from traders from Tripoli. The bishop reached for the heavy clasp at the centre. He fed a key into the lock and turned it. Slowly he raised the lid. Then he bent down and lifted out an umbrella of rare beauty. It was topped by a cross, as on the roof of a church, and so laden with gold and silver that when he opened it out – kwa! – it trembled and rang. 'Take this,' he said, turning the handle toward her. 'It's yours.' And then he was gone.

When she woke she asked her husband what the dream could mean.

You will give birth to a son, he replied. And he will be a shade and a protection in the world.

'AND BEHOLD THERE WAS A CERTAIN WOMAN TRAVELLING WITH THE COMPANY WHO WAS WITH CHILD, AND HER TIME FOR BRINGING FORTH WAS NIGH, AND SHE WAS UNABLE TO RUN AWAY WITH THE OTHER PEOPLE. AND SHE CRIED OUT AFTER THOSE WHO HAD FORSAKEN HER AND FLED, BUT NO MAN TURNED BACK TO HAVE REGARD UNTO HER, AND SHE FOUND NONE TO HELP HER, AND DESPAIRED UTTERLY OF OBTAINING HELP FROM MAN.'

She had been labouring for two days already, and the priests had been reading in turn, from the Dirsanè Ruphael, and now from the Miracles of Mary.

'AND IT CAME TO PASS THAT, WHEN THOSE WHO HAD TAKEN FLIGHT ARRIVED AT THE SEASHORE, SHE STRETCHED OUT HER HANDS, AND RAISED HER EYES TO GOD IN HEAVEN, AND MADE

SUPPLICATION TO OUR LADY MARY WITH GREAT OUTCRY AND
WITH MUCH WEEPING AND LAMENTATION.'

Her husband, appalled by her labour, had remained in the room
– the receiving room, this time; so many people were attending
this birth there was no space for them in the main house. He
bowed to the ground in prayer and wept so much he asked for a
cup to be brought, a dark cup, hollowed out of cow horn. His tears
dripped into the rough bottom and when he had collected a
finger's height he handed it to one of the women. Give her this to
drink. Maybe it will hasten the birth.

Mariam Mariam Mariam Mariam Mariam, dirèshilin.
O Lady Lady Lady Lady Lady, come to our aid.

As dusk drew in on the third day she slipped, exhausted, into a
state between sleeping and waking. And in that state it seemed to
her that she had gently taken leave of herself, but was watching
herself at the same time, that she lay quiet on her side near the fire,
and that a young girl approached her, a girl of about six years old.
The child was beautiful. She had an oval face, with perfect skin
and a small straight nose, full lips, and tumbling, glossy hair.
Silver chains spilled down the front of her long dress. The child
came close and stretching out a small hand stroked her belly and
her straining back. And it seemed as though her loins responded
to the child's touch, easing and calming, and to the child's voice,
which felt as if it came both from outside her, and from within her
own heart. 'Ayzosh. You will be safely delivered.' But when she
turned to give her thanks the child was gone.

'THEN WHILST SHE WAS IN THE MIDST OF THE SEA THE PAINS OF
CHILDBIRTH TOOK HOLD UPON HER, AND OUR LADY MARY TOOK
THE CHILD FROM HER WOMB; AND SHE GAVE BIRTH TO A FINE
BOY. AND HIS MOTHER CALLED HER BOY "ABRASKIROSPAS"
[WHICH MEANS] IN GREEK AND HEBREW "THE HAND OF MARY
TOUCHED HIM AND BLESSED HIM IN THE WOMB OF HIS MOTHER".
NOW NEITHER PAIN NOR FLOW OF BLOOD CAME TO HIS MOTHER.'

At midnight, in the third night after the third day, the baby, a boy, was born.

He was completely silent. His eyes were closed, and there seemed to be no life in him. Memories of the last death twisted through her and she cried to the midwife, Madam, have I laboured and laboured in vain?

'Quiet,' answered the midwife, sharply. Then, more kindly, 'Ayzosh. He's probably just tired.'

A big bowl was filled with water, and soap brought, and clean clothes to receive the child, and the room stretched taut with watching. As soon as he touched the water he heaved a great sigh and began to suck his fingers. And it was as if her spirit flowed back into her body, as if she had suddenly come back to life herself.

'AND IT CAME TO PASS THAT, WHEN THE SEA RETURNED TO ITS
OWN PLACE, AND THE WATERS THEREOF BECAME QUIET, AND
THE WAVES WENT DOWN, THE WOMAN WENT FORTH FROM IT
CARRYING HER CHILD IN HER ARMS. AND WHEN THE PEOPLE
SAW HER MANY OF THEM MARVELLED AND BECAME SPEECHLESS
BY REASON OF THIS GREAT AND MIRACULOUS THING.'

For three days after her son's birth she slept on the floor, on a thin pallet surrounded by strewn grass and tracked-in rainy season mud, welcoming the cold wind that gusted under the door in the darkest hour before dawn, offering her body and her comfort to Mary, who had heard her in her greatest distress.

When forty days had passed and he was taken to be baptised she would name her first son Edemariam, or hand of Mary, but in the meantime she reached for him and held him close. Looking down at the flattened curls of wet dark hair, the breathing, fragile skull, she knew suddenly that this too was a kind of salvation; that these small forms that emerged from her buffeted body might be an answer to her loneliness, the depth of which she was only now beginning to comprehend.

The sun was just beginning to touch the tops of the furthest mountains when they set out, but the city was still in shade, the air clear and chill. She tightened her arms around the baby, and shifted her weight. Breastfeeding had further stripped her already slight figure, and the bright cloth decorating her mule did nothing to soften the saddle. The sound of their mules' hooves – hers, Alemitu's, his – echoed against the walls of the houses, but their servants' bare feet, trotting alongside, made no sound at all. Above them the curved swords of eucalyptus leaves soughed in the breeze, bowed, crossed each other, bowed again.

She leaned forward and grasped the pommel as they began to climb, and watched as the sun slid down the slopes to meet them. It lit the tops of the trees and picked out the straw and pebbles in the mud walls of the houses, which glinted it back. Along the roadside the grass had dried to feathery fringes of pale gold, cool in the dapple of early morning. Woodsmoke rose tentative into the air, and crows argued themselves hoarse over rubbish heaps. Women stood in doorways, beating basketwork clean with branches. Hens scratched at their feet, and cockerels. Her mule's ears twitched, pointed forward again.

Houses with doorways open onto the street began to give way to homesteads encircled by fences of lashed-together eucalyptus and euphorbia. Flowering bushes crowded in on them, wild roses, creamy yellow crotons. They began to pass women in white shawls returning from church, stepping easy, quiet among the rocks, their long shadows mingling and moving together. When the women saw her husband they came close to ask for his blessing, touching forehead and chin to the cross he disentangled from the folds of the gabi he had wrapped around his shoulders for warmth. May God bless you and keep you. Amen.

At the sides of the path thistle flowers, white and purple starbursts nestled in green-pointed ruffs, drew level with the tops of the eucalyptus trees on the hillside below. Lammergeyers wheeled, then sloped down out of sight. The far ridges of the mountains were grey-blue steps ascending into a sky undisturbed by any clouds at all. The mules' hooves crunched against the rubble on the path and the sound seemed suddenly smaller, bare, but also hard and bright, as though it could travel forever through the clear air above the valley. As they rounded a bend she turned slightly, and saw Gondar spread out below them.

Ever since the beginning of the dry season, as the ground hardened, the green meadows began to yellow, as the rivers shrank and became passable, she had felt the city change around her. Her husband did not think to tell her much about the wider world, but she saw and heard enough. She knew that in the market there were more people, more strangers, sensed a darker, harsher mood. The servants came in and out with water and wood and shreds of news. Thousands of men and women, their mules, their children,

their slaves, were walking in daily from the mountains. They
carried muskets, spears, shields, lion's mane headdresses; grinding
stones bent the women's backs, and great hide-covered food
baskets chafed the donkeys' flanks. At night the mead-houses rang
with war chants, with boasting and with burnished memory. The
emperor's cousin Ras Kassa, appointed governor of Begemdir and
Semien after Ras Gugsa Wulé's death, was calling his armies in.

Kassa was as pious as Ras Gugsa had been, and known for his
mastery of theology, but though he had fought in the battle of
Adwa as a teenager and been victorious against Negus Mikael of
Wollo since, he was not necessarily known for his mastery of war.
He was steady and loyal, and a trusted adviser to the emperor –
even though he had a better claim to the throne. When he became
governor he had preferred to stay at the court in Addis Ababa and
delegated the administration of Gondar to the eldest of his four
sons. She would hear, decades later, that on one of her husband's
trips to Addis he had been charged with a message for the new
emperor: Gondar and the provinces of which it was capital were
too important to be treated in this way. He had been seriously
heard, apparently, but for years felt resented by Ras Kassa's sons.

They were travelling between scrubby fields now, scattered with
yellow stones, the occasional bush a dark jewel set in dry gold
land. Now the path was runnelled and gullied, scoured and scored
by the daily deluges of the rainy season, and the mules slowed,
picking their way along tracks that narrowed sometimes to a
single hoof's width, a steep drop on one side, rough drystone wall
or the long unforgiving thorns of an acacia on the other. Or they
cast, wary, around the occasional darker section of ground, where
the earth was still soggy and pockmarked by previous traffic. Step
there and the likelihood was a broken leg, and the mules knew

that as well as their riders. Often the animals paused, thinking for a second or two, scanning the ground ahead before stepping on. They sighed, huge, gusty sighs. Their mouths dripped, and their haunches were glossed with sweat that soaked through her dress and warmed her calves and thighs.

Gondar dropped out of sight.

The town had emptied of people as abruptly as it had filled, and for a few weeks had felt quiet but stretched out of shape, waiting, but uncertain what it was waiting for.

And then one day an answer: six specks in the sky, specks moving faster and straighter than any bird, growing bigger and bigger, until she could hear them roar.

Oh mother of God, what is this? Snatching up her daughter, the baby, looking frantic about for somewhere to hide. Oh daughter of David, save us.

Closer and closer the specks came. They looked like crosses now, stubby dark crosses, trailing smoke. The streets ran with women, children, clergy, the infirm – anyone able-bodied had marched away with Ras Kassa or quietly disappeared. As the thundering drew near they threw themselves into ditches, huddled against walls, behind trees. Oh Queen of Heaven, save us.

Around again. She didn't see but was soon told how on the second pass, over the castles, a dark rain fell from them, a hail of metal that exploded with a terrible noise as it hit the ground. How many huts caught fire, and the women and children inside them.

That was when the order came from the emperor, who when the Italians invaded, marching over the border and finally taking Adwa, had headquartered at Dessié: evacuate Gondar during daylight hours, every day.

*

So here they were, travelling away from Gondar, as they had travelled yesterday, as they would travel tomorrow, and the day after
that. They had crested a long rise and were looking down toward
the Shinta river. Vegetable plots had been planted along its banks,
and neat rows of silver-green kale rose up the slopes. They picked
their way down to the water and dismounted under a stand of
bayberry trees. The mules' necks shivered, and their tails swished
at the heavy flies. Most of the river had shrunk to mud, but a
small stream still trickled through the main channel. Green algae
waved around rocks like hair in a breeze. White and yellow
butterflies flicked above the water; the mules' muzzles dipped
down, then away again, disdainful. Distant children called to
cattle.

When the mules were rested they began to climb again, into
rich farmland in harvest time. Everywhere pale gold domes of teff
waited to be threshed, peas and beans hung heavy on vines, plots
of glossy green chillis, of kale and tomato, bustled in toward each
other. Men trotted past, staffs supporting on their heads piles of
straw nearly as high as themselves. The ground grew steeper and
rockier as the riverbed fell away. The mules' mouths dripped.

A circle of dark trees crowned the hill, and within the circle
stood a low church. Tethering the mules outside in the shade, they
stepped under the acacia guarding the entrance. It was cool and
calm inside the perimeter walls, under junipers and olive trees so
old they towered above the church's thatched roof. Dry leaves
cracked underfoot, turtledoves cooed, bees buzzed around a hive.
She bowed, made the sign of the cross. Made her way forward and
bowed again, so her forehead touched the walls of the church,
then her lips, forehead, lips.

Gonderoch Mariam, which her husband had led for eight years
now, was old, far older than many churches in Gondar. Thirteenth-
century, said the more historically inclined priests. Originally

named Debrè-Genet, or Mount of Paradise, by the king who built it, it had long since been rechristened Gonderoch because so many people from Gondar were moved to walk up the mountain, to pray under its kind trees. Often when she and Tsega came here the younger deacons and priests would drop to the ground, making to kiss her husband's feet. He would bend before they could do so, cupping their faces, raising them up, presenting his cross instead.

As they left the church compound she looked across the valley, back toward Gondar. The Italian aeroplanes had not merely terrified her, and her children and her neighbours; they had underscored, emphatically, that while churches, and especially churches at this height, had been always places of safety, they were safe no longer. She thought of their nearly two-hour-long treks back to Gondar in the evenings. Their occasional sightings of hyena droppings. The night she and the children had watched her husband climb up and shove his old rifle deep into the thatch. The day the dark dots had appeared again. She had never felt such fear, fear that was a kind of pure pain, which tightened her chest and loosened her bowels so she had to run behind a tree to empty them. For endless minutes they hardly breathed, waiting for the explosions. But they did not come, and the plane continued on its way.

Sometimes they did not go back down the mountain but stayed overnight with a patron of her husband's, who owned land all around the church and had settled some on him. They could see it from her doorway: seven terraces divided by low stone walls dropping in wide steps down from the house, which was on the north-eastern brow of the mountain. Barley grew here. From the bottom of the terrace stretched the flat summit, shared with other landowners and planted with teff. Past the church, down the rocky slopes toward the Shinta, was more barley, of which they also received a share. Beyond the fields she could see the river valley, and Gondar, and then, in the hazed distance, the mountains that surrounded Gondar.

The dry season wore on. The farmers brought their tithes down the mountain. Wild figs darkened in the trees. The peaches mellowed, the potatoes and tomatoes, the gourds and chickpeas and peas ripened and were harvested. In the hollows jasmine bloomed. The caravans rattled through the markets, rushing before the rains. And with them came news. Ras Kassa had engaged the Italians at Tembien, in the mountains between Gondar and eastern Eritrea, but had been unable to push them back. Two months later he had been forced out of those peaks altogether. There were whispers about sheets of rain falling from aeroplanes, rain that stripped and burned and blinded, that dripped from the bushes and poisoned the lakes, that sent even the very bravest fleeing. The emperor gathering around him a vast and growing army.

We must join him, said her husband. I will not fight, being a priest, but I can minister to the soldiers. You will come with me. You will not blow trumpets or clean rifles or sharpen swords, as the wives of soldiers do, but like them you will bring your servants, and cook food. Gather provisions and prepare the household.

She bowed, yes, and did as she was told, struggling to seem calm despite her fear. But the days passed, and they did not leave. She watched as priests came and went, served them as they talked, listened and watched, and eventually gathered that many did not agree with him. He was their leader, they argued. Surely he should stay, to protect them?

And so they were still at Gonderoch Mariam when the Italians entered Gondar. Dejazmatch Ayalew Birru, Ras Kassa's son-in-law, had declined to engage in guerrilla tactics that might have stalled them, and the foreigners had simply walked in, hundreds and hundreds of them, following vehicle after vehicle.

The small rains began. The giddying smell of fresh-wetted earth rose from the fields and, despite everything, ploughing had to begin.

The news grew worse and worse.

The emperor was defeated. Routed. Listening aghast they could almost see the terrible disordered retreat, the mountain passes jammed with fleeing men and women. Imagined the aeroplanes chugging through the air above, the fire they dealt, the terrible unnatural rain. The vultures circling across the day and the hyenas laughing through the night.

And then, at first almost incomprehensible – because wasn't that one of the most fundamental expectations of a leader, that he should stay with his people, and die on the battlefield if need be? – the knowledge that Emperor Hailè Selassie had fled. He had taken his family and boarded a train to Djibouti. Ras Kassa and his youngest son had left for Europe too. Three days later, Italian columns entered Addis.

In Gondar the foreigners began building at once. Roads, hotels, banks. A rifle factory and two hospitals. An airstrip. An avenue of shops and hotels unfurling southward, down toward the castles. Water began to flow through narrow metal pipes, and at night the streets were strung with small glowing orbs. The city rang with the sound of hammers and chisels wielded by local workmen on higher wages than they had ever seen before. And in more than one mind grew the unspoken question – or if it was spoken, uttered only at home, looking about, over shoulders and across rooms – who else ever did such things for us?

Some days after the Italians entered, another deputation of priests, leaders of some of the forty-four churches, climbed the mountain. The foreigners have asked to see you, they said to her husband.

Have they.

You are our leader, and they have asked to see you.

In that first meeting ranks of foreign soldiers in tight-collared black shirts and shining black boots faced ranks of priests in their whitest turbans and brightest embroidered robes. The Italian leader, a red plume dropping full over an eagle perched on the front of his hat, declared Victor Emmanuel III of Italy emperor of Ethiopia and required all present to sign a statement acceding to that fact. After the signing, after the priests had danced and the uncomprehending foreigners had smiled, her husband had stepped forward and requested that the Italians return to the churches all rights and lands they had confiscated. But the request was either misunderstood – the translator's street Italian being no match for a churchman's careful perorations – or simply ignored. For not only did the foreigners not return the land, they took churches for garrisons and moved into the homes of evicted or suddenly absent aristocrats.

Some castles became offices, others headquarters for the carabinieri. A smooth dark road poured past the hotels, the new cinema, the shops, but stopped abruptly at the castles; below that, where the city continued into the Saturday market and the bluffs overlooking the Qeha, everything was still bare earth. Electricity stopped there too, and piped water. For a long time she did not notice. Now that they were back in the city everything for her remained as it had always been, sweet water available from her well, the market outside healthy and bustling and equal to her needs. Even the talk of deliberate division – whites here, locals there, a school and a hospital and a courthouse each, no locals allowed in cinemas at all – made little impression on her. Gondar had always been a divided city, between Muslim, Christian and Jewish quarters, between aristocracy, gentry, artisans and peasants, and who wanted anything to do with these foreigners anyway? It was they who often insisted on crossing over, chasing women, living with them, defying orders from their superiors.

Other orders were more efficiently upheld. The death penalty for men caught possessing arms, for instance, a quixotic aim in a land where bearing arms was a necessary adjunct to any claim to be taken seriously, where manliness and honour were synonymous with physical courage and the willingness to go to war. Summary executions, then, often on testimony of nothing more than a rival with a grudge; imprisonment and flogging of families which refused to give up weapons or state the whereabouts of those who carried them. Terrified whispers of a portable gallows, dragged from village to village. Of decapitated heads held high. The death penalty for anyone suspected of supporting the absent emperor.

Her husband shut his face, and with a gift of a thousand Maria Theresa thalers he had received from Emperor Hailè Selassie before the war, with the income from Ba'ata's holy water, the tithes from Bisnit and Dembiya, the income from church arbitration, market dues, the sale of the gold-filigreed capes and robes Empress Zewditu had given him on his first promotion, set about building his church as if the end against which it was spiritual insurance could arrive any day.

The inner circle, the holy of holies, was held together with mud, trampled by labourers' feet over and over for three or four days, but the meqdes, the priests' domain, was to be constructed of stone only, cut so carefully no mortar would be needed. All day new-quarried pink tufa arrived from Qusquam, a quiet hilltop north-west of the city, carried between pairs of former slaves, or on the backs of donkeys. All day, in between prayers and sermons and confessions, between the endless questions and supplications visited upon an administrator of forty-four churches, he wove among the labourers, correcting a cut of rock here, a misunderstanding about size there, cajoling, ordering, threatening, driving the work as fast as he knew how.

For he saw, moving about the city, how in the Italians the need to prove superiority over those they had vanquished, and increasingly over their *fear* of those they had vanquished, had resulted in an overreaching brutality. Fear of the nobles and village elders, whom they relieved of their positions, replacing them with Italian or Eritrean mercenaries, and not infrequently sending them 'to Rome' – bundling them into cars and aeroplanes and from thence to either prison or death. And beyond all this a kind of ancient dream-fear, too, of the Orthodox Church's ancillaries: its deacons and monks, its soothsayers and its wild-haired travelling hermits, who looked on the surface to have little power but transmitted information faster, it sometimes seemed, than any telephone.

They feared the priests too. Or, at least, were deeply suspicious of them – though in that they were not unusual. Even amongst their own colleagues and parishioners priests often had a venal reputation, of being concerned more with status and possessions than with matters holy; of being inveterate, individualist schemers. For in the way that the emperor had total power over every aspect of his subjects' lives, priests had power over their spiritual weather. They received confessions, they punished and they forgave, they controlled access to the written word and thus to the Bible and all its interpretations. To this was added, through tithes, the possibility of worldly riches, and even more temptation. The Italians saw this power and its possible uses (openness to influence, a source of spies) – they also saw that priests were either unreliable, or an active, potent threat. Both sides had only to think of the days after the fall of Addis Ababa, when it quickly became clear that by shooting Abunè Petros, bishop of eastern Ethiopia, the Italians had created a martyr.

Exactly a year after the Italians first bombed Gondar, they hunted down and shot Ras Kassa's eldest son. Two of his younger brothers were lured into submission ten days later. They had been promised safety but were promptly executed.

And in Gondar her husband was again ordered to bring his priests to the main square.

Not far from the huge sycamore fig stood two clerics, facing Italian guns. The leader of the church of Gana Yohannes stood still, the priest beside him babbled and shook. We grew up together, we were children together, will we die together? But the aleqa of Gana Yohannes said nothing, and then the friend of his childhood said nothing either.

When he returned to her she thought, this is how the dead must look. His face was like soot. He did not seem to see her. For two full weeks he could not be persuaded to eat.

She had been at it for a while, chopping the ginger and garlic, mixing it with cardamom and basil and rue, stirring it through simmering butter. The sun was warm on her head and on the baby sleeping in the shawl on her back. She glanced over at her eldest daughter, sitting in the doorway of the house. Her worries about how her children would look had not applied to this child, at least. Alemitu, six years old now, already had a nice long nose and wide brow, a graceful neck.

Are you hungry?

She took a spoonful of the freshly spiced butter and, mixing it with some berberé, poured it over a piece of injera, soaking it and tearing it into rich bite-size pieces. Here. It will make you grow.

The afternoon wore on. The sun seemed, if anything, hotter. Sounds receded. The corners of her daughter's mouth glistened. She kept working.

The next time she looked over she was at her daughter's side almost in the same motion. Hands like startled butterflies, loosening the neck of the child's dress, feeling her face, which burned, a small dark sun. Cradling her, calling her name. Feeling it in every sliver of herself when Alemitu's body snapped rigid as a hide left out to dry. Her chin flung back. White eyes stared at the sky. Oh Mary mother of God, what is it? What is it?

Bring her clothes! A shawl! But her husband had just returned from a trip to Addis Ababa, and everything was down at the river with the menservants, being beaten clean. There was only a thin muslin veil with which to cover her daughter and lift her into the cool of the house.

Go get her father.

When he came, he had a friend with him, and the two men exchanged fierce whispers over the child's inert body. She must take holy water, said her husband. 'Holy water won't work.' Only the devil could do something like this to her, look at her body, how stiff and contorted it is. She must be taken to holy water immediately. 'It won't work.' Yes, it will.

Her husband prevailed.

A neighbour offered to help, and every morning, for two times seven days, the small group set off into the dawn, heading for the little old church of Teklè-Haimanot. For two times seven days they sat amongst all the other supplicants, waiting their turn. So much sadness in the world, she thought, looking at the array of bodies before her. So much care. The stripped flanks of farmers accustomed to sparing food and abundant labour. The much-suckled breasts hanging flat and soft. The warped and twisted young limbs. The torsos shining with wellbeing, their specific curses invisible. The underdresses sticking to bodies dripping, bodies drying, bodies inward-looking under the sun. And above them all the perfunctory deacons crouching, pouring the blessed water, and then, as midday approached, intoning – one eye on the takings and another on lunch – the acts of the saints and the Miracles of Mary.

On one of these days her neighbour invited her to stay over-night, so they could go to the church together in the morning. She was a member of an association that met once a month and it was her turn as host; Yetemegnu could sit and chat for a while if she wished, before she went to bed. She accepted, and made sure to prepare all the food her husband would need the next day.

But when she made to go he said, you are not leaving.

Why? They are like family, nothing will happen to me there.

Stay here.

She knew what he wanted, but pretended not to. I must go –

Stay here!

In the silence after the pain she felt the blood tip, cool, down her forehead. It dripped off her eyebrow and traced the side of her nose.

She turned, picked up her children, and left.

He sent a deputation of elders. 'Come, you must be friends,' they said. 'All marriages have their difficulties.' It was a metal belt buckle! He could have blinded me! 'In that he was not right, we will tell him he was not right. For your part, you know what it says in the book, that women must obey their husbands, as Sarah obeyed Abraham, calling him Lord.' But he could have blinded me! 'He will not do it again.' She was tight with protest and anger, but she did as she was told.

Eventually Alemitu's fever ebbed, but it was as if the child had gone too. Previously curious and high-spirited, she was silent and withdrawn. Sullen, her mother thought, as yet another comment or request for help went unheeded. She watched everyone and everything as though from across a plain, as though she were not really there.

One day Yetemegnu took Alemitu on a visit to her aunt Tirunesh, and when she had to return to look after her husband, left her in the older woman's care. It was some time before she pieced together what then happened. Tirunesh, accustomed as she was to unquestioning obedience, had become increasingly impatient. She would say Alemitu's name and there would be no answer. 'Come here, child.' No reply. 'Sit there.' Nothing. One afternoon one of her brother's tenant farmers, a Muslim from Simano, came to visit and she regaled him with the problem. 'The parents are nice enough, but this girl, quite rude.' 'Madam,' he said, 'you do understand she's deaf. The holy water has made her deaf.'

She was at Ba'ata when she heard, and the crows rose out of the junipers and flapped and cawed at her screaming. And then she was running, pulling at her dress, her girdle, tearing them with her nails, ripping the soft fabric, scratching at her face, crying erri, erri, erri, hoarse with bitterness and grief, crying as if someone had died.

She ran to the house and snatched up her daughter. She could not, would not go home until she was healed. From church to church, from Ba'ata to Medhané-Alem, from John Son of Thunder to Gabriel, to Ruphael, to St George. Stumbling from spring to spring, with this awkward bundle in her arms, or tied to her back like a baby. Presenting her to priest after priest, begging them to touch her ears with the Gospel. Oh Lord, open her ears. Oh God in heaven, hear my prayer.

To the springs at Wanzayé, which steamed into the morning. Three days there, receiving water that flowed so hot from the ground it was said King Woldè-Giorgis once had a piece of meat thrown into it that came out cooked. Three days, praying that water people said had once opened a grown woman's ears would do the same for her daughter. Three days of easing into small enclosures, where sweat dripped off their faces and stung her healing brow, and three nights sleeping in the open, drawing her shemma over and under them, shivering as the temperature dropped, listening to the rustle of sheep against eucalyptus and the occasional call of a jackal. And on one of those nights, a deep, sharp pain in her leg. When they rose in the morning the limb had swollen to twice its normal size.

They tied it tight, hoping to prevent the spider's poison from going up any further, then when that did not work they wrapped her in a gabi and carried her home. For three months she dipped between fever and waking, and waking would call out her daughter's name, to be answered only with silence.

It had been a difficult pregnancy, and now it was a difficult labour. She did not scream, but she could not suppress a bitten-down hum.

Hmmmm.

Behind the curtain her husband prayed.

Hmmmm.

The women held her tight across her chest as she crouched and strained.

Mariam Mariam Mariam Mariam Mariam dirèshilin.
O Lady Lady Lady Lady Lady, come to our aid.

Hmm. Hmmmmmm.

Tiny wrinkled feet, and a complete change in the quality of attention from the women surrounding her.

Hmmmmmmmm.

Hips.

Hmmmm. Hmmmm. Hmmmm.

Finally, finally, a head. Quickly the women lifted the baby, took soot from the fire and anointed its face with a cross.

Twelve ililta. Another boy.

The exhaustion lasted for weeks. One morning her husband came to her having been up all night cooking, as carefully as he knew how, a freshly slaughtered sheep. Gently he sat down beside her, tried to feed her the watery sauce, picking out pieces of meat so overcooked they had fallen into strands. Eat, my heart, eat. But she kept shaking her head, all she wanted was sleep, could he not let her sleep?

Losing patience, he called her aunt, who laughed. She can't eat something cooked like that! But even a day's worth of remedial spicing and simmering could not tempt her, and eventually all the neighbours who had assisted at the birth were invited to feast instead.

The child thrived. His hair curled and his face filled, and he basked in the attention of a household that knew him to be especially precious, because of the manner of his coming. When forty days had passed they christened him Yohannes, after the feast day on which he was born.

Her husband was away when the woman came. The palm branches were brittle over the lintel, and a sheep fossicked in the yard.

'Where is he?'

The lack of preliminaries, a serious sharpness, startled her out of her diffidence. She looked up. In Belesa with the priests. Talking to the shifta.

The woman, a friend of the family, nodded. It was an increasingly common species of trip for senior clergy, whom the Italians were sending into the countryside, into Mamara and Qimant and Quoleña country, with orders to make contact with the patriots (or bandits, or shifta, depending on who was describing them, and to whom). Tell them, they said, to return to the city. Persuade them not to fight us. She didn't think her husband did any persuading, or that anyone ever came back with him. But for the moment he went as he was told.

For things were changing. From Addis Ababa came accounts of grenades thrown, missing their Italian target. And of the viceroy's three days of revenge: streets full of running, of blood and fire. Of bodies thrown over bridges or dragged behind lorries. Of women raped, disembowelled, of children torched in their homes. Of monks shot en masse. Of a cathedral, and a monastery, burning. By the end of that rainy season the new Roman province of Amhara, so easily taken and so far so peacefully ruled, was blooming with revolt: revolt overt, in the form of ragtag warriors massing in the surrounding hills, and revolt covert, in the form of people like the woman before her, a comfortable, kind lady who was trusted by the Italians, but who everyone knew worked against them too.

The viceroy was recalled, and replaced by one with somewhat subtler tactics. The Duke of Aosta had seen the pride with which the Ethiopian church claimed its five bishops, and now saw also how it could be used. Of the five, Abunè Petros was dead, as was Abunè Mikael. Abunè Saurios had fled to Jerusalem, and Abunè Isaac was in prison. Only Abunè Abraham, who served flamboyant, insurrectionary Ras Hailu of Gojjam, had followed his temporal master and formally submitted. Aosta offered full separation from Egypt, with Abunè Abraham as archbishop. And Abunè Abraham accepted.

Oh the fuss the priests made when their new prelate arrived in Gondar by aeroplane shortly afterwards. And the numbers who had listened to his sermon! 'Gondar is famed for its scholars, for its population by you, the learned. And yet you stand by and tolerate the impertinence of the farmers at Armachiho, who live so close to you! Why do you not advise them to stop fighting? Should you not be spending your hours thinking about how to repay the Italians, who have done so much for our church?' When he had finished, and was turning away, her husband spoke. 'The best time to influence the farmers is after the harvest. Then they might even be persuaded to become monks, let alone to stop fighting.'

All the next morning Abunè Abraham toured the market, reiterating his message.

'It is a sin to refuse this government God has bestowed upon us, this government that has freed our church. If you accept the foreigners, you will gain riches on earth and inherit the kingdom of heaven. But if you do not listen to these orders from me, your spiritual father, you will be punished on earth then consigned to hellfire.' Again and again, working himself up into righteous paroxysms, until finally the Italian colonel who had brought him took him to the new airstrip, from thence to Gojjam, and finally back to Addis Ababa, where he further proclaimed the government would protect the churches, rebuild those that had been destroyed, and restore property and estates.

Not long after, her husband, like all the other priests and imams, began receiving a salary, 10,000 lira, which he added to the remains of the imperial gifts, the rent, the sale of livestock, and spent on the church. He hired Italian workmen. Although she didn't cook for them – she thought pasta looked like hookworm and was baffled by galeta, eventually grinding the biscuits with spices and mixing them with butter to make a surprisingly tasty paste for her own family – she sent servants out to them with eggs and goat meat from Gonderoch Mariam, so that they would eat well, and finish faster. Occasionally they got a sheep, too.

They were obviously skilled, and useful, but she was afraid of these red-faced men who made themselves at home so easily, and looked at her so assessingly. One morning a foreign man had wandered into their compound.

'Anchi!' You! The familiar form, rather than the respectful form which was her due. 'Any eggs?'

Non ché, she replied, reaching for something that sounded like no.

'Niq niq ché!' he replied. A rude gesture.

What? You donkey! Get out. Get out!

If she had had a stone within reach, she would have thrown it. Not that it would necessarily have been a deterrent. He returned often when she was alone – except for the servants and children, of course – each time sending her rushing about the house, locking the doors, then running up the stairs to peer at him from a window.

Another day she needed some logs split, and accompanied her servants to the church to ensure it was efficiently done. The Italian foreman had grabbed her small hand in both of his, and begun to stroke it. She snatched it back.

Min abatih! You bastard! What do you think you're doing?

'Anchi buono,' he replied. 'Anchi buono.'

Much later she would laugh about it, but then what she mostly felt was an almost physical wave of danger. There were so many stories. How at dawn the roads around the garrisons filled with women going home. How queues led to brothels demarcated by colour: yellow for officers, green for soldiers, black for local troops. But also how many Ethiopian women lived as wives with Italian men. (And how a few of these men were imprisoned for declaring love, thereby breaking Italian law, which allowed only for sexual necessity.) Only the other morning she had come down the stairs to an unwonted silence in the kitchen hut, and shadows on the floor in the corner where the servant girl used to sleep. She had had to grind her own teff ever since.

They had tempted and spoiled so many clergymen's wives. And what if her husband, eyes permanently alive to the possibility of betrayal, jealous, it sometimes seemed to her, of the very air, what if her husband saw? It wasn't that long since they had had an important guest for lunch. She had thought she had served everything necessary, that all was as it should be, but when she left the room, she had felt her husband behind her. Had felt his presence, and then cold metal, pressing hard into the back of her neck. She hissed a breath in through her teeth. He pressed harder, and she understood he was using the base of his prayer stick. Did you not realise you could have killed me? she asked afterwards, when the guest was gone and the dregs of fear had turned to anger and goaded her into speech. I would have died and you would have gone to prison. Who would have raised all these children then?

Come, he said, another time. Come where? She had had her hair braided, had bathed and put on her best clothes, but that was because she loved the feel of cleanliness, the smell and softness of fresh-washed cotton, not because she expected to go anywhere. She had wound Yohannes in another length of clean white muslin and tied him to her back, where her breath and beating heart had carried him into sleep. Come. Let's have a picture of you. She laughed, disbelieving. It's true. There's a photographer at a house down the road. Let's go.

The photographer was an Italian. Her husband looked back and forth between the foreigner, who fussed and tweaked and clicked, and his wife, who sat straight and still. In the photograph her skin is clear and youthful, tight over a long, perfect jawline, but she seems older than twenty-two or twenty-three. Her eyes look out as if from a place of safety somewhere behind her head. A suggestion of a frown shades the inner edge of each brow. For the rest of her life she would look at this photograph and think, if I had known what would happen, if I had been more sophisticated, I would have held my son in my arms, and we would have faced the camera together. But it did not occur to her, and after the Italian had finished her husband took her home.

But mostly she worked. She fed all the Ethiopian labourers – fed the hewers of wood and the hewers of stone and the carpenters and the masons who joined what they had hewn together: fifteen people, sometimes twenty, including one strong man who could eat three injera at a sitting. Fed the four servants who helped her, fed her family. Fed any guests her husband saw fit to bring home. Fed any relatives who walked in from the countryside or across the city and sat in the receiving room, silent or relaying news, pausing to rise and bow whenever she or her husband entered. Fed them roasted grain and dark beer at lunchtime, and then a full supper in the evening, when she stood and circled and watched and made sure everyone ate as much as they were able. The servants cleaned and ground grain, picked through raw spices, drew water, but it was she who baked huge piles of injera every day, and cooked the sauces, who boiled coffee and sat on a low stool pouring it. It was obvious to her that her husband had no idea how much labour he was creating when he brought yet another man home and said, he's been working, let him eat, no idea even though he saw that she rose at five every morning and did not untie the thick girdle that supported her back and held up her growing belly until after eleven at night. Carefully she would sit down onto the hide bed and ease herself flat, listening to the finally quiet house, thinking, this church is killing me.

So now she looked at the woman in front of her, knew, from her tone, this was no routine visit, and fear licked through her like flame. What is it?

'He is in danger. You are both in danger. A gun has been hidden in the church, in the vault below the holy of holies. They know about it. They are coming to search your house. They are coming now. Don't be scared. There is no time to be scared. We must search everything before they come, to make sure nothing has been hidden here as well.'

And so they began. Under the seats in the receiving room, under the hollow bases of the mesobs, Edemariam, who thought it all a good game, toddling busily about behind them. Through the clay pots and sacks in the hut where the servants ground grain. In the rafters. Behind the piles of cooking pots in the kitchen hut, inside them, under them, while a Maundy Thursday porridge of wheat flour and split broad beans plopped over the fire. In the storehouse on the ground floor of the main building they sifted through the madigas of teff, of dry peas, of lentils. They plunged their arms into the mess of fermenting red wheat waiting to be made into beer for Easter Sunday. They swirled their hands through the dregs of previous beer-makings, and through the water pots. She felt behind the clothes hanging on hooks in the bedroom, rifled through all the sheets, under the children's blankets, under the servants', into every possible cranny. Nothing. Then she remembered their own gun in the thatch, and shoved it deeper, as deep as she could reach.

'Good,' said the informer, and took her leave.

And, just as quickly, it seemed, was back, alongside the carabiniere.

How handsome he was! Shining black boots. Belt fastened tight. Buttons like new thalers. Gun held just so. She could feel herself shaking, praying it wouldn't show.

She could hardly bear to watch as he followed their trail around the house. After a while she became aware of a movement in the corner of her eye. She glanced at her son. He was reaching for the ceiling, pointing up to the thatch. She snatched him up, pulled in the fat arm. He wriggled free, and reached again. She held him more firmly, and he fought her, squalling.

She glanced at the carabiniere. Had he noticed? But he was dragging his hands through the grains, peering behind the large clay amphorae. Rolling up a sleeve and plunging an entire arm into a gan full of pea-water and dregs, where he paused so long her heart skittered with panic. Had he found something? What had they missed? Her shaking intensified till she thought she might fall. But his hand came up empty. He shook the water off, fastidious, and finally, finally, he and the informer left.

It was late on the day of the Crucifixion when her husband returned, and she washed his feet and put food in front of him, and waited, bowed, till he had eaten before she told him.

Master. Her voice like someone else's. Someone has planted a gun in the church so the foreigners think it is yours and kill you.

You must disappear. Go, now, to Gonderoch Mariam, and the Qimant will escort you to Metemma. There are patriots there, fighting for Hailè Selassie. Join them.

No.

Go. I will stay here, with the children. They won't touch me, and if they do, it's God's will. Go now.

No.

They will kill you. I cannot watch you die.

I will not go.

Have you not spent these last years watching your colleagues die? Have you forgotten the priests executed in front of you while you prayed for their souls, fasted for them, grieved for them? Would you have the same done unto you? I beg you, go.

The Mother of God hasn't abandoned me yet. I will not leave.

When darkness fell that Good Friday he wrapped himself in his cape and gabi and stepped out through the gateway. It was mid-morning on Saturday when he returned.

He's dead.

Who?

The man. The man who planted the gun.

What?

All night, he said to her, while around him his priests and deacons chanted out their books of hours, as they read from the Old Testament and from the New, as they recited the Miracles of Mary, he had prayed, standing, kneeling, bowing until his forehead touched the rough ground. Oh my mother, please do not let me die for something I have not done. Oh my mother, I beg of you, keep me safe.

After the dawn and the service of peace, after the priests had blessed and handed rushes to each worshipper to tie about their foreheads in memory of the grass the dove brought to Noah, he had made his way slowly out of the little church, past the junipers and olive trees, the drunkenly leaning gravestones. Slowly he became aware of a woman weaving among them, a young slip of a woman like his wife, keening for someone dead. He recognised her.

Who died?

My brother, came the answer. He was found in his bed last night. No gunshot wounds, no evidence of poison. Just dead.

In the house near the market he looked at his wife. Yetemegnu. The Mother of God killed him for me. He tried to kill me, but she heard my prayer.

And he had gone straight back into the church and bowed and bowed again, praying his thanks.

Now she watched as he took a cape embroidered with gold and drew it over his shoulders, as he wrapped a silk turban around his head. She watched as he picked up his prayer stick and fly-whisk and left for the Piassa and the castles, where all the priests and deacons from all the forty-four churches of Gondar, and the rural churches of Begemdir and Semien – though now it was called Amhara – all the men the Italians had asked him to bring, milled about the walls. They carried heavy gold crosses and swung gold censers; they sheltered under umbrellas of red and green and blue silk, they held aloft yellow wax tapers. Their capes blazed with golden suns. So many standing under the sycamore fig the earth was hidden, they told her, as far as the eye could see.

The foreigners were there too, in their finery. Man after man from both sides climbed up to the balconies of Empress Mintiwab's palace and addressed the crowds, vying to outdo each other in panegyric ardour, priests buying favour for favour's sake, pomp and position being pomp and position and particularly material gain whatever its provenance, but watching also for emollient effect, aware always – had they not had plenty of practice, under emperor after warlord after emperor? – that present flattery bought precious things: fragile safety, time, space. Space in which to decide whether and when and how to act, and thus to wrest fragments of freedom and power from a time in which both were in short supply. And so they called out to their flocks, Look! Look what the Italian has done for Ethiopia and Ethiopia's church. Look how the Italian respects our great religion, and increases the power of our faith.

But the Italians had their ruses too, one of which was a belated decision not to allow themselves to be used as instruments for the settling of private scores. And so one of the highest ranking among them beckoned to Aleqa Tsega. 'Come here,' he said, to the slight dark administrator of all these hundreds of men. And, as she heard it later, he reached forward and clasped her husband's hands. 'I know what happened this morning, how you were saved. You must be a very holy man.' And he raised his voice and addressed the crowd. 'From now on I will hear nothing, no rumours, no spies, nothing against Aleqa Tsega. I will believe no ill of him.'

So when the cock crowed before dawn on Easter morning and the long hard fast was broken with crushed flaxseed and honey and drumbeats and ululation, when the sheep's blood ran in rivulets and reddened the green grass in the yard, when the gans of barley beer and honey wine were poured out and when their guests, flushed and loud with holiday plenty, sang her their poems of praise – Oh brewer of mead, Oh brewer of mead, wash your hands in honey – when even on the second day the priests faced each other in rows and shook their sistra and danced, there was an extra freight to their celebration.

*

She had never been able to face food when she was tired, and now the exhaustion of the work was exacerbated by a pregnancy that made her so sick she could not eat even if she had tried. Her cheekbones clawed out of her face. Her arms and legs were knobbled sticks. She grew so thin the large belly she carried before her began to look obscene, added on.

Her husband watched with concern, knowing from her previous lyings-in how specific her requirements would be when she finally gave birth, how the rough sauces of the well-meaning village women would go untouched. So he took some fresh beef, cut it into even strips, and hung it on twine strung across one of the storerooms. When it had curled and dried into a dark red jerky he took it down and placed it in a leather bag, ready for when she needed it.

It's for my child, he said. It's for my child, for when she gives birth.

The labour this time was fine, though after the last one almost anything would have seemed so. She named the boy Teklé-mariam, plant of Mary, and a gelded sheep was slaughtered, but the message had not got out and there were few guests. Her husband called a neighbour to him and handed her the leather bag. Take this. Break it up in a mortar and cook it with spices and butter and pieces of injera, just as she likes it.

When it was ready he brought it steaming in to her. Slowly, trying not to disturb the baby in her arms, she sat up and took a grateful mouthful.

Meetings, then, meetings and more meetings. All the necessary meetings of an aleqa and a liqè-kahinat – the questions, the applications, the letters to dictate, the promotions and demotions, the importunings and the blandishments, the straggle of petitioners anyone with any power at all carried with them in their daily round. And he, as administrator of forty-four churches, had plenty of power. But even to her, to whom he told nothing, it was obvious the number of meetings was increasing.

Meetings, meetings, meetings, so his movements became entirely unpredictable, and she instantly regretted the one instance she took advantage of an absence to leave the house. She had been preparing shirro, dried peas and chickpeas pounded fine with cardamom, fenugreek and rue, when she realised she had no sieve. She rocked the new baby to sleep, picked up her skirts and ran to the neighbour's. She didn't go in, but stood at the doorway, asked for the sieve, ran back. She had hardly entered her home when she heard a thud next to her shoulder. She turned. A machete was embedded in the door; the thick wood had split, above and below the blade.

She had given her husband one wide look and left as fast as she had come. And the elders had been furious with him. What if you had killed her? What in God's name were you thinking? What has this child ever done to you? That time she had stayed with her relatives for weeks.

Meetings in the refurbished castle of Empress Mintiwab, with the Italian governor and his deputies, in which, standing in front of a fellow priest about to be hanged, he argued that if they must punish to punish fairly; that if unfair sentences such as this went ahead, all of his priests would remain in the public square, in protest. The priest lived.

Meetings with his foreman, meetings with masons, meetings with the two painters he had imported from Gojjam, who were covering the walls with faces, with robes of blue and red and green and gold: Pious Gelawdewos, defeating Ahmed the Left-Handed at Zantera in Infiraz and saving the Ethiopian empire for Christianity; St George forever slaying his dragon; Satan horned and bound in flames; Mary and her son, guarded by archangels.

More meetings in the wide space outside the castle walls, huge meetings, where the Italians indulged their taste for mass rallies, for appropriated imperial pomp and for grandstanding speeches, where the priests responded in kind, and everyone watched everyone else like vultures. 'I will keep my mouth,' said her husband, walking down the massed ranks of his priests and deacons, speaking low and urgent, quoting the psalm in the old language only they could speak. 'I will keep my mouth.' None of them would be here unless they had learned the rest of it by heart: 'I said, I will take heed to my ways, that I sin not with my tongue: I will keep my mouth with a bridle, while the wicked is before me. I was dumb with silence, I held my peace, even from good; and my sorrow was stirred.'

Big meetings, disguised as parties, for as his church grew, so did the circle he made for its influence. The clergy were used to celebrating the yearly feast of Ba'ata with handfuls of roasted grain and a horn or two of dark beer. But on the third day of the fourth month, when the heavy rains were a memory and the lush green of the valleys was giving way to brown, he sent invitations not just to the usual clerics, but to other dignitaries too: church administrators, the headmen of outlying villages, the old warrior Ayalew Birru. Aged Abunè Abraham. Come, Tsega said to them all, help me to celebrate Her Presentation and Entry into the Tabernacle. Come to my house, there is food waiting. They arrived by the dozen to eat the fasting stews she had prepared, and he walked among them, making sure no one went hungry, that the servants kept the drinking horns full to slopping over. Each year the feast continued for longer, guests arriving not just on the third of the month, but for days after.

Small meetings, to which she was also witness, with their family confessor and his son, who slipped in after dark and wrote letter after letter for a husband who paced in and out of the trembling pool of lamplight, dictating their contents because his father's curse still held and he was not allowed to write.

Meetings, and more travel, setting off among groups of convivial, bowing, chronically suspicious clergy, leaving at dawn in a clatter of mule hooves and clanking bridles, the deacons half-walking, half-running alongside, Tsega inward-facing and stern under his large white turban, the letters his confessor had written tucked deep into his clothing. Official meetings with churchmen, with elders, but also, afterwards and in secret, with sharp-eyed men hung with belts of bullets, whose hair had been left to grow long and matted and stuck out from their heads like lumpy thatch, who listened to his reports of Italian movements, took the proffered papers, read them, then buried them in their own clothing, ready to be passed on. Meetings at churches across the wide province, licensed by his job, by his aleqa-ship of Gonderoch Mariam, by the Italians, to range free through valleys and mountain passes steadily slipping out of their grasp.

Meetings of which she knew so little that when one of his trips extended into weeks she panicked, believing the Italians had taken him, as they had taken so many others, to 'Rome', and she would never see him again. She wept until her neighbours said, Yetemegnu, you must stop. You are watched, we are all watched. The Italians will hear of it. But she could not stop. So they said, go. Take your children into the countryside and go. And she lifted Teklé onto her back, and they all walked up the mountain to Gonderoch Mariam.

After a month he returned. Abunè Abraham had died, and the prior of Debrè Libanos had been promoted into his place; the Italians had indeed flown her husband north – but to Asmara, to participate with colleagues from the other provinces in the choosing of a phalanx of new bishops. On the way back he had visited the holy city of Aksum and used some of the money the Italians had given him to buy a high rich curtain for Ba'ata's holy of holies. The party to welcome him back lasted days.

Meetings, meetings, meetings – and then one day a silence that made her look up. He was holding a piece of paper in one hand and a dense dark object in the other. Of course she could not read the writing, and he did not explain. But she could see the double circle stamped in ink at the bottom of the paper, that within the circle strode the muscular marching body of the imperial lion.

When the food began to run out the Italians could no longer deny that the hills crawled with guerrillas who watched the roads and owned the villages and backed them into the dusty lanes of Gondar, where they began to protest against their own leaders. Five small pieces of bread had cost a lira when the Italians arrived five years ago; they now cost fifty. A quintal of teff, once ten or fifteen lira, was rising toward two thousand. The markets flapped with Italian uniforms, sold for cabbage or for grain. When she heard the foreign soldiers had been seen eyeing donkeys and mules, considering what flavour of meat they might provide, her fear of them began to shade into what she one day recognised was pity.

For months the city had simmered with rumours. That Hailè Selassie might return from exile in England after all. That Italy had declared war on Britain, its ally in the Horn of Africa, and the British government had suddenly seen its way to helping an Ethiopian emperor to whom it had given safe harbour but otherwise frustrated and ignored. That Hailè Selassie, after five years' exile, had travelled to Khartoum. That Ras Kassa and his only surviving son had left Jerusalem to join their emperor. That British troops were organising in the Sudan. That together all had crossed the border and entered Gojjam.

Her husband redoubled his absences, preaching in the villages, in the hamlets, in the open country; preaching his monarch's message of qualified forgiveness, hope, ultimatum (no mention of regret, however, and certainly no apology): 'My people, lacking me, went to the enemy,' ran the lion-sealed letter. 'Let them know that We are coming. Let them know that their land and arms will not be touched if they return to me. Muslim, Falasha, Qimant, Raya-Azebo, Adel, Denqala or Somali – all are sons of Ethiopia. Tell them to arise. Persuade them to fight for Us.' And many of these sons were, with belated vigour, doing as they were told.

One night, in the chill hours before a clear dry-season dawn drew open the receding veils of stars, her husband left their compound, made his way to the new church, and stepped inside. The painters had done well. The canvases glowed in the taper-light. Behind the biblical stories – and one panel in which he himself appeared, paler-skinned and splendid in a cape of indigo trimmed with gold – were walls so strong, stones so tight-joined they could, he hoped, withstand anything. It was a pity he had not yet been able to build the final, outer circle, but the stone had been quarried, and lay in heaps under the trees.

After the vigil and the benediction, after communion, he and his priests emerged blinking into the sunshine, and he saw they were waiting under the trees as promised: the Italian deputy, and Yohannes the new archbishop, white-bearded and severe under a burnished turban. The priests had danced: bare feet on now-warm earth, eyes holding eyes, so they sank and rose as one, sank and rose until another priest appeared, a flat weight covered in rich cloth balanced on his head. Ilililil! cried the women as the tabot,

Ba'ata's replica of the Ark of the Covenant, began its first circuit, echoing the diurnal round of the sun. Ililililil! as it was taken into the new building and placed on the new altar, consecrating it. After the archbishop's speech, after the thanks directed by the archbishop at the Italian deputy, after the Italian had left, Tsega stood at the top of the clean new steps. Come, he said to all before him. Come and help me to celebrate this feast of Her Presentation and Entry into the Tabernacle. Come to my house, there is food waiting.

Two weeks after Emperor Hailè Selassie regained Addis Ababa, Gondar acquired a new Italian governor, who took office just as the small rains began and the farmers started to turn and soften the earth, preparing it for seeding. Guglielmo Nasi made it clear he would expand the policy already tested by his predecessors: trust local leaders, or give the impression of trusting them, respect custom and due process, and they will reward you with loyalty – or, more likely, the impression of loyalty. And he had had some success, so that when the old Shoan general Birru Woldè-Gabriel, reputed to be an illegitimate son of Emperor Menelik, returned from Jerusalem and camped with a British detachment at Denqez, many Gondarés were wary. They had heard the British were more zealous than the Italians when it came to exclusions and conde-scensions based on skin colour – why invite that upon themselves? Furthermore, the British might be hailed as saviours now, but it was only a few months since they had been the opposite. Rather the devil they knew, given the choice. Some, waiting to see how the balance of power would tip, even smuggled food into the city at night.

Not that it helped much. Nasi minted coins and printed bank-notes near the airport at Azezo, but the only difference anyone could see was they had to use larger and larger piles of cash for scarcer and scarcer goods. The pack animals were slaughtered and made into stews. She heard some soldiers had even tried to eat grass. She heard they wept with hunger, and she wept too, taken by a kind of rushing grief for young men far from their mothers, lost and afraid in a country that was starving them into submission.

The patriots, emboldened, gave out titles and fiefdoms as rewards and incentives to fight. And Nasi, watching intently, learning fast that in this country office and rank were potent currency, did the same. Legions of men from all stations of life were called to the castles and awarded titles by this worried-looking little man whose eagle-embossed cap was an overwrought weight resting on round spectacles and a very sharp nose. On one day alone he awarded 110 titles, from generals to aleqas, commanders of the vanguard to lord chief justices. Another day, it was said, he stopped to watch some farmers harvesting teff with sickles in the fields. After a while he ordered them to divide into three groups. This group, he announced, are henceforth Commanders of the Right. This group, Commanders of the Left. And this group, Commanders of the Fort. Ayalew Birru became a ras, a duke and head of the army. And her husband was named bitwoded, or chief of the counsellors and beloved of the realm.

As for her, little changed until one day he raised a stick again.

She spoke before she could think: Raise your head. Let me have a good look at you.

The rod paused, quivering just beyond her vision. At last he said, What do you want to see?

She did not answer, but met his eyes, steady.

The silence rose and spread. At last, when she felt she could be still no longer, his arm dropped, and he turned away.

AND WHILST WE WERE IN JUDAH THE OLD MAN, JOSEPH
THE CARPENTER, MINISTERED TO US AND FED US. AND
BEHOLD THE ANGEL GABRIEL APPEARED UNTO HIM,
SAYING, 'RISE UP, TAKE THE CHILD AND HIS MOTHER,
AND DEPART TO THE LAND OF EGYPT, FOR THEY ARE
SEEKING FOR THE CHILD TO DESTROY HIM.' ... AND WE
WERE ON THE ROAD TO DEBRÈ KUSKUAM. NOW WHEN
WE ARRIVED THERE WE WENT HITHER AND THITHER SO
THAT WE MIGHT PERCHANCE FIND A GOOD PLACE
WHEREIN WE MIGHT DWELL ...

*– THE HISTORY OF THE VIRGIN MARY RELATED BY
TIMOTHY, PATRIARCH OF ALEXANDRIA*

And then one day aeroplanes roared back into the skies above
Gondar. They were followed by a thick white rain that resolved
itself, on the ground, on the roofs, in the trees, into sheets of
paper. Those who could read explained to those who could not –
'We are coming,' it says, the emperor is coming, the planes are
British planes – but she refused to believe in friendship that
arrived in such a way. Every time the planes returned the fear
incapacitated her, and they returned so often she became ill, so
that at last her husband took her back to Gonderoch Mariam. But
here too the dark deliberate wings passed overhead.

Eventually, early one morning, her husband and a cousin and
a manservant lifted her, pregnant again, onto a mule. She stretched
out her arms for the baby and waited as mules and donkeys were
loaded with the older children, with clothes and cooking pots;
with a hundredweight of the whitest teff, with forty kilos of dried
spiced chilli and a generous measure of shirro. Her husband
kissed each child goodbye. When they finally set off, a single
manservant trotted alongside.

The sun was high when they came to a scattering of neat thatched huts. A couple of cows stared at them, and a busy goat or two; chickens scratched, and the bustling headman-tax collector hovered and bowed and ensured the little group knew they had the best possible beds, and the best possible food. After a few days they moved on, picking their way between fields of tender green chickpeas, golden teff, sorghum, millet, maize, over earth by turns red, brown, and a deep rich black. They were in Dembiya now, a wide plain, famously fertile, that dropped gradually toward Lake Tana. Emperors had holidayed here between military campaigns, or used this as a base from which to launch them; Ahmed the Left-Handed had built a home here at the height of his jihad; caravans of Muslim slave traders had crawled through during the dry seasons; monasteries were established, palaces, churches, everyone in their pleasure and greed attempting to ignore the region's great drawback: malaria. The little party climbed up scrubby slopes, past high euphorbia trees, each thick-ribbed and scalloped branch tipped with incongruously delicate blossom, up to the brow of a low hill and the hamlet of Atakilt Giorgis, where her father came out to meet her.

Gondar had been attacked in earnest, he said. British planes were raining bombs, not paper now, on the airport, on the mountain passes, and on the anti-aircraft guns guarding the north of the city. The Italians had sent up aircraft of their own, but these had been shot out of the sky, so when, every day, the air raid sirens howled out their warning, Italians and locals alike ran to hide in holes dug into the ground. Many new buildings were in partial ruins and Iyasu's castle had suffered a direct hit. And her husband had gone to join the soldiers of Dejazmatch Birru, now engaged at Denqez and at Degoma.

And so in Atakilt Giorgis she settled down to make a home. A small home, unlike the one she had become used to, but satisfactory – curved mud walls, a packed-earth floor, a thatched roof, a bit of land, all for only thirty coins of silver. As the situation worsened her husband had converted all their money from lira back to thalers. Now she took the bag of remaining silver and buried it, making sure to scratch out the evidence.

Atakilt Giorgis was a beautiful place, even though in his cups her father swore up and down that while circumstances might dictate he had to live here for a while, he could never be buried among the animals in such a wilderness. If she stood in her new doorway and looked back the way she had come she could see the plains, and then, on the horizon, the long high plateau, flat as a tabot. Before her the land sloped down to a small river, Qench Wiha, and then, close-dotted with simiza bushes and acacia trees, leisurely up again. To her left, on a slight rise, stood the vast sycamore fig under which the farmers met to discuss seed prices and cattle sales, or for long raucous parties. Sometimes they saw monkeys leaping through the tree's sturdy branches, or, underneath it, earth fresh-turned by rootling aardvarks. At night they all heard the squeaking, whistling cries of jackals, and the unhinged chuckles of hyenas. Once she saw an unfamiliar figure standing under the tree. A sun-darkened, wind-cracked face. Hair a wild abundance of matted points. Clothes white at some point, but now a muddy shade of grey. Sleeves tight to the elbow, jodhpurs ending in naked ground-broadened feet. A wide belt of dull gold bullets, and a rifle nearly as long as he was tall. Quickly she retreated into the house.

Her father had remarried and she knew she had to tread with care. She gave freely of the food she had brought from Gondar, and when one of her half-brothers arrived trailing donkeys loaded with spices and more teff and two full madigas of fixings for barley beer, she gave of those too. The gifts were eagerly received, but when they were finished, she felt they were not reciprocated in kind. Used to the finest grain and knowing the bounty of the surrounding countryside, she was quietly dismayed by the rough bread her new stepmother provided, by the flimsy tales of storehouses filled with infected crops.

One day she was folding clothes into a wooden chest when her stepmother accosted her. 'Who do you think you are? Why are you throwing the gifts my mother gave me onto the ground?' She looked up. Her stepmother was holding a short necklace of black silk and gold filigree. 'Why are you using this chest, anyway? Did you come to inherit this place and everything in it?'

As the birth of her sixth baby approached she became aware that her father saw the tension between the two women and was trying to protect her. He directed a trusted servant to override his wife and give his daughter and her children only the best of food, butter and milk. She said little, not wishing to draw attention, but made sure he knew she was grateful.

Mariam Mariam Mariam Mariam Mariam, dirèshilin.
O Lady Lady Lady Lady Lady, come to our aid.

'Yetemegnu,' said the maidservant. 'Strong beer is good in labour. Drink a bit of this.' She didn't like beer, but she assented. The woman poured her a full horn. After a while she accepted another.

Mariam Mariam Mariam Mariam Mariam, dirèshilin.
O Lady Lady Lady Lady Lady, come to our aid.

She was on her knees when her fourth son arrived, so big that her father exclaimed in surprise that he had not killed her, but so fast, in comparison to the other births, that it seemed only a moment before Molla was stretched out, lusty and yowling, in the midwife's arms.

A couple of days later she was woken by a commotion in the yard and walked gingerly to the doorway to see what was going on. A lunch of greens gurgled on the fire and over it her father and stepmother stood facing each other, stark with rage. She had thought her father was at a funeral in a neighbouring village; obviously he was back early, and just as obviously had taken beer. Her stepmother had been to market for fresh butter and, it transpired later, had stopped off at a friend's for a measure of araqi, leaving herself no time to add spices to the pot.

'How dare you! Did you think I wouldn't notice?'

'Oh, go away, you slave of a consul!' his wife snapped back. 'You corpse!'

He snatched a smouldering stick from the edge of the fire and cracked his wife across the head with it. Blood bloomed from the gash and flowed down her face.

Yetemegnu rushed forward, objecting, but her father turned, making to hit her too. She screamed, the new baby wailed, the other children cried, and in the commotion she did not at first see her stepmother's older son advancing, rifle cocked.

She rushed forward again. No! Kill me instead!

'Sit down, girl,' said her father. 'Go and sit down!'

He dashed into his house and brought out his own gun, by which time half the neighbourhood was in attendance, and the two men were forcibly disarmed.

Back straight, head held high, her stepmother walked out of the compound, sowing blood as she went.

The next few weeks were quiet and in many ways contented. She recovered from the birth and began to cook for her father and for her stepmother's children as well as her own. Sometimes she walked to the little round church perched on a crest of land above the plain, where the calm was only emphasised by the number of birds that hooped and chattered through the olive trees and fern pines, and in the euphorbia planted to mark the graves. Red ants scuttled through the couch grass. Yellow butterflies turned and darted. She would kiss the little arched doorways, feeling how worn they were, how softened by age and veneration, and slowly make her way back to the house. Often she found herself thinking about her stepmother, and eventually asked her father if she might return. 'It's up to her,' came the gruff response. 'I didn't send her away.'

She went to the village elders and asked them to intervene. The devil has entered our lives. Please help us. Finally, on the morning of her son's christening, her stepmother walked back into the compound and took up her duties as if nothing had ever happened.

Yetemegnu only saw the patriots a couple more times, but she knew they were about. Everyone talked about them, all the time, with an indistinguishable mixture of thrill, admiration, disapproval and fear. Often with an amused understanding, too, that while many really did fight for their country, or for their country and for revenge, the war had also given inveterate brigands a cause with which to ennoble ongoing feuds and the hunt for personal gain.

Over coffee, or while they picked through grain, cleaning it, the women told stories – about famous brigands of the past: of Shiguté, who with his two-hundred-strong band had ruled the lands west of Lake Tana; or of Tewodros and Yohannes, brigands first and emperors after. They dropped their voices and spoke of brigands present: 'Did you hear who's joined the fighters around here?' 'That's no surprise. He always had that tendency. But he's kind, unlike their leader.' 'Yes – their leader is dark, and cruel. He steals cattle and grain. He abducts women.' They turned toward her. 'Come to think of it, someone overheard them talking the other day, and he's interested in you.' Her heart skipped. 'Yes, you. He says he likes the light one' – that was her stepsister, who had also just given birth – 'and the little dark one.'

After that she was terrified whenever she heard they were near. She knew what they would have seen: that she glowed with youth, and with the richnesses of recent motherhood, and finally she sent a message to her husband on the battlefield. Lord, I am in danger. The cord that was tied about my neck at my christening, binding me to a blameless life, is about to be snapped. They are going to kidnap me. Come quickly.

He came at once.

She was shocked when she saw him. He was slumped toward the neck of his mule, exhausted, unwell. She watched anxiously as he helped her to gather up her household, directing the servants in their packing, selling the little thatched hut, working out what they could and could not carry; she would always remember they had to leave all their teff behind. They sent Edemariam to a monastery so he could continue his education. And then they mounted their mules again and clopped down onto the open plain.

They travelled for hours, through fields and up rocky slopes, picking a way through forests and splashing through streams, aiming, like so many other refugees from Gondar, for Denqez.

After a while she began to notice the absences. Every house they passed was empty, the hamlets silent. Fresh graves crowded the churchyards, and dogs slunk intent among them. When they arrived in Denqez, the watchman tried to turn them away.

Let us in, don't you know who we are?

'I don't know and I don't care. You can't stay here.'

He's a priest. Show some respect!

'Shut up, you.'

Her husband sighed, and out of his shemma drew a letter with a seal, this time Dejazmatch Birru's. Impatient, the watchman examined it; brusque, he waved them through.

That night they stayed in a house from which they could tell the family had recently fled. Exhausted, she slept.

And she dreamed – of the road they had just travelled, and of a crowd of women in their finest white dresses. They shaded their faces with bright umbrellas, and walked lightly, joyfully, as though it were a holiday. As she looked back at them first one and then another raised their voices and called, ilililili! ilililili! until all of them were ululating, high and continuous, and her head rang with their trilling and she wanted to hide or to stop up her ears but could not.

They're diseases, she said to him when she woke. I know they're diseases. And they're happy, because we've brought them with us.

It began as a lassitude, a greyness of countenance and spirit. The children stopped playing, they stopped talking, and worst of all they stopped eating. She made them barley porridge, dripping with butter, but they would not touch it. She made honey-laced gruel, because it was easier to slip between their lips, but it dribbled back out again.

She sat by them, the new baby strapped tight inside the front of her dress so the diseases would not know he was there, and prayed, desperately. Oh my Lady, help us please.

They knew no one in this land, she felt there was no one she could call on. She crushed wild olive bark into a powder and put it in water for them to drink, she rubbed their bodies with healing leaves, but none of her medicines made any difference. Then her husband, already ailing, worsened. He drank a horn of water, and his urine was blood.

Every morning Alemitu, who was still well, helped her clean the vomit of the night before from the younger children's mouths. They swept the floor and strewed it with fresh grass to lighten the air. Every noon the fever took hold of her sons and shook them as if it would never let go. And at night it visited her husband, whirling him into deliriums in which he chanted out blank-eyed liturgies. When he woke he would ask how she had slept, and she could not answer, well, by the grace of God, as she would usually have done. Instead she would say, I could not sleep, I kept calling your name and you didn't answer; it was as though I had become a plaything of the spirits, and I did not know what to do.

Three months after they had begun the fevers eased, and Teklé and her husband began to sit up, to look about and talk, and to eat. Spring warmed the ground and dried out the riverbanks and the hut filled with heavy black flies, bumping around the patients' heads, settling at the corners of their mouths. Her husband was well enough to eat by then, but he would not eat during the day, because of the flies.

But Yohannes, her beautiful Yohannes, who they had said looked like her father's father, who had almost killed her when he was born, he did not rally. He lay in his bed, getting steadily weaker, his stomach growing into a taut mound under the gabi while the rest of him wasted away.

One morning she told her husband she had had a terrible dream and she knew this son would not survive. He looked at her for a long moment and replied, I dreamt that too.

Because the church of Mary in Denqez was too far for her to walk to, she had formed the habit of going out into the yard and, bowing her head in its general direction, praying for their health in the open air. This time her prayer was different.

Oh Mary, mother of God, I place my trust in you. If this child must die, please may he die at home, in my own country, where there are kin to bury him, and where he can rest in the church that we have made. Oh Queen of Heaven, I implore you. Please ensure we do not have to leave him here alone. Amen.

And she walked slowly back into the house.

BOOK III
1942–1953

TAHSAS
THE FOURTH MONTH

Summer. Hot and sunny with occasional light rains.
Red aloes and thorny plants dominate scorched landscape.
Rivers become brooks, brooks go dry. Main harvests
finished. Peasants take teff to market and seed summer
barley in irrigated fields. Traders load caravans with
produce and hides to exchange for salt and imports;
cattle pastured in the valleys. Judges begin riding circuits
to hear litigation.

When finally they heard it was safe to return they gathered up their children, their animals and their belongings and walked over the hills into Gondar. They avoided the Saturday market. Their house had been squatted by Sudanese soldiers, long gone, who had sold the corrugated iron from the roof and torn out the doors for firewood; a family they knew huddled in the lower rooms, and their bedroom yawned at the sky. They avoided the church, too. It belonged, now, to a priest who had gone to see Emperor Hailè Selassie in the Sudan, a canny petitioner whose name meant 'he stole the light'.

On, up through the city, past the royal enclosure. Over the next couple of months, they would hear about the battle for Gondar, which had turned out to be the last major battle for Ethiopia. Thousands of warriors had stormed down the mountains, careering over minefields that had accidentally been cleared by a herd of stampeding cattle, and taken back their city in a tumult of fire and drunkenness. The Italians had retreated, with all the city's petrol, into the castle grounds, then spent an anxious night lobbing the burning flares thrown at them back over the walls.

Away from the centre, toward the north-west, and into the neighbourhood now called Otto Barco because that was where the Italians had kept their cars. As they retook the city, patriot leaders had claimed and then begun to parcel out choice areas and buildings; her new neighbours were military men and minor princelings, and their temporary home a six-room villa with an indoor toilet and running water. There were still streetlamps in the Italian quarter, and in the evenings she marvelled at their steady light, at the impression they gave of advancement and of order.

But the markets were full of weapons, of rifles and grenades, deliberately left by the retreating foreigners so order would be the last thing possible in the city they had hoped would become a showpiece of their empire. The slave trade had begun again: day after day there were stories of children disappearing and traders proliferating, picking off the darker-skinned, promising marriage to naïve young women, then selling them in the border towns, in Welqait, under the fig trees in the north-west corner of the Saturday market. Selling them for thalers, or bartering them for guns.

There was very little food, and so the returning exiles cooked together, each providing what they could – one family donating a clay pot of shirro, another of hot meat stew. She had brought with her from Denqez a madiga of teff, and another of wheat flour, and so she provided wheat bread, pancakes, and injera.

Indoors, however, Yohannes could not eat. The glossy hair she used to cut with such care was rough and dusty-looking, and the whites of his eyes had yellowed. Holy water did not help; the hospital for locals was a ruin, that for the Italians damaged. So she sat by him and held his hands and kissed him, and washed his face, which memory would make the most beautiful of all her children, and prayed over him, to the guardian angels, to Mary, to Mary's son. She sat by him in the days and through the nights, listening to his shallow breath against the dark, or to the sudden rifle shots and shouts from the mead-houses. The fights over women or plunder or allegiance-in-hindsight: who resisted for all five years, who for three, who for a few months, who resisted not at all.

The city was a scavengers' playground. Vehicles littered the streets. Many had already been gutted for parts, but others still functioned, and in the daytime her older children, Edemariam now back among them, took to climbing up into abandoned lorry cabins, turning keys in their ignitions, driving the hulks forward for a few yards, then leaping out and running away, giddy with transgression. She herself kept an eye out for barrels and for sheets of corrugated iron. She knew how useful these things would be, for beer-making, for water-collection, for mending their roof, and when the overworked police tried to stop her, she ran to their confessor to borrow money she could slip into the policemen's hands, persuading them to turn away and allow her to remove her booty in peace. In this way they rebuilt their own home, and were in such a hurry to return they moved in when only one big room was habitable.

But still the fevers took Yohannes, shaking his body, engulfing him in sweat, sinking him so far into exhaustion that he could respond to nothing; taking him again and again and again. His breathing became laboured, hoarse and irregular, and then one day Edemariam, sitting at the foot of the bed watching his older brother struggle, realised that he was still.

At once they took her away, and shut up the room, and held her as she rocked and keened. He's well, he's well, they said, he's in safe hands, he's well, but she would not be calmed, because she knew that well meant dead, that she was being held here because a mother cannot be allowed to see the corpse of her child, that what they were doing in that room was washing the small tormented body, wrapping it up tight, closing its eyes.

In Gondar the custom was that the dead were waked for seven days. But after they had buried Yohannes in a quiet corner of Ba'ata the receiving room filled with mourners for a full fortnight. Those who had been in exile, those who had gone into hiding all over the city, people they had not seen for years came bearing food and sat to weep with them.

She felt her own grief as a pain that stopped her breath and raked through her like a kind of insanity, but she was horrified by her husband's sadness, frightened by it. He would shut himself in their bedroom, and when she tiptoed to the door, worried he had not eaten, she heard the harsh sobs of someone unaccustomed to crying, and turned away. When he came out he was dark and bitter, as though he had been scorched.

I thought he would be my brother, he said to her once. That he would be my country and my shield. One of her half-brothers took her aside. 'This man is going to die on you,' he said. 'I saw my father die of grief like this. You need to stop grieving yourself, and take him in hand.'

So she opened the door and watched him for a moment, and, remembering her own mother, said to him, Ayzoh, grief dulls as time goes on. But he did not seem to hear. So she spoke again, Please stop. I'll die if this continues.

The next time she went in he said, We must have another child. Remember we have all these children already, she said – Alemitu, and Edemariam, and Teklé, and Molla. But yes, we'll have another. Please stop.

Eventually his tears eased, and, outwardly at least, the self-contained man she was accustomed to was restored, and returned to his work.

So when the crown prince, round of jaw, bulbous of eye and, being fresh from military training, upright of bearing, came with the British forces to take official charge of Gondar, her husband was there to welcome him. And after the ceremonial retaking of the castles, after the bowing and the ululation, Tsega stood again at the head of his priests, a black-caped liqè-kahinat, to incant in Ge'ez the necessary qiné of blandishment and fealty.

The sky may fall or the earth rise
But David's son Hailè Selassie will not be dethroned.
The fertile vine of Menelik will bear the fruit of
    wisdom
For the Italians have been weakened, and are caught
    in a snare.

A long road ran before her, cutting through the countryside, through irrigation green and insistent against the dry brown land. The road ran before her, narrowing, until it faded to a mote at the horizon. She was being pulled along it, faster and faster, her feet barely touching the ground, her right arm outstretched, and at the end of her right arm – Yohannes, holding her hand and running before her, bright-eyed as he had been before the fevers took him, and when she saw him she cried out.

And woke. I'm going to die! I'm going to die before this baby is born.

Ayzosh, ayzosh. What is it?

So she told him, sobbing, and he said, Ayzosh. The long road means you'll enjoy a long life. That's your age. It's not death. Ayzosh.

At first she was calmed, but over the next days she could not shake the foreboding. He encouraged her to fast, but still she was unassuaged, and on the third day she went to him and said, I'm going to mass.

How can you do that? We have no servants, what will we eat? I will go to Mary for you, and tell her your fears.

But she said no, no, and the next day rose early to bake the injera and cook the greens and split-pea sauce of fasting season, and was finished just in time to hook her shemma over her head, climb onto a mule, and clop up to the Church of the Treasury of Mary, Gimja-bet Mariam, a little old building set into the castle walls. She was so pregnant she couldn't stand for long, so she sank to her knees and prayed there, twisting and turning on the floor in search of a comfortable position, until mid-afternoon, when the service was over and she mounted the mule again and re-entering her own home took her first meal of the day.

She did this for two or three days, but then she thought, if I go on a mule, in luxury, Mary will not listen to me, so the next morning she walked up the hill.

After a week her confessor, who had been watching her prostrations and how often they were accompanied by weeping, asked the priests of Gimja-bet Mariam to go into their sanctuary and to bring out one of their treasures. 'Here,' he said, placing before her a picture of Mary, 'this was painted with the tears of Emperor Yohannes the Just. Cry to this, and the Virgin will hear you.'

She did as he said, and toward the end of the two weeks another dream came to her. She was standing in a building with a woman who looked like one of her aunts. The woman turned and eased herself through a tiny window, and, once outside, beckoned to her, come, come to me. But she could see only windows, no doors, and she said, I can't, look, my belly is too big. The woman kept beckoning, come, come, and finally she put her head through the window, and then her shoulders, and suddenly she was outside too, in clear air. The woman held out a length of bamboo. She grasped it and followed, walking until they came to a deep fast-flowing river. She was terrified of water, had always been terrified of water, and she refused to go on, but the woman said, ayzosh, step where I step, and nothing will happen to you. So she followed, placing trembling feet on one stone at a time. Once on the other side they walked until they came to a bubbling spring in a wide meadow, a hot spring that steamed into grass covered with tiny white flowers and touched by the wind. The woman said, this is your place. And she sat her down in the quiet and left her.

In the morning, when she woke, she knew the danger had passed. She told her husband what she had seen, and he said, the meadows are the world, and the water is sustenance. You will live for many years. You will have many children, and you will see much.

The sun came out when her seventh child, Tiruworq, was born.

Sometimes now, in the calm after supper, when the low basket-work tables were cleared away and her husband, or increasingly Edemariam, had said the closing blessing, she told stories.

Have you heard the one about the partridge and the rooster? And they would turn, expectant, her husband too sometimes, if he was home. All right.

The partridge and the rooster became friends. First the partridge invited the rooster to her house in a lush green tree. It was beautiful, there was no smoke or dark, just great hospitality, and he spent the night. The next morning when they woke the rooster said, let me show you my home now, and the partridge agreed, and off they went together. He gave her food and water and was chatting away when he realised the partridge wasn't answering. He turned to remonstrate, but stopped mid-sentence. She had been hung by her neck from the rafters. Oh, the sorrow of that discovery! And for ever after the rooster has cried qo-qo-oh!, qo-qo-oh!

Another story. There was once a hermit, a holy man who had left the world to travel alone through the wilderness. He had no grain to eat, no bread, but every nightfall, wherever he had travelled and whatever time of year it happened to be, he would find a partridge caught in a snare. And he'd give thanks for her, cut her throat, pull out her innards, wash her, then roast her over the fire, and that would be his food for the day. One morning he met two hermits who had been travelling for years. They walked on together, but all the while he was worrying – what shall I feed them? One partridge isn't enough for all of us. And so they walked and walked, and when night fell they found not one, but three partridges in a snare. They gave thanks and began to cut their throats, peel off their crowns of feathers, pull out the innards, ready to cook them over a fire of sweet wood from the forest and share out the meat. But as they were doing this one of the new hermits said, 'But it's a fasting day! We can't eat meat.' In an instant two of the partridges rose into the air and flew away. The hermit realised what he had done and fell to his knees, sobbing, 'We didn't know! We didn't know! We didn't know it was a gift from the Lord. Oh, we have insulted Him!' They bowed down and begged His forgiveness, and then they rose and returned to their homes.

More usually she would feed the children, put them to bed, and wait for him. She had said she would not eat until he came home, and he had said he would do the same for her. But it was he who was often late. And when they finally sat down he would delay them further, wrapping sauce and injera into big mouthfuls and carrying them into the room where his children slept, waking and trying to feed the sleepy forms, forcing mouthfuls on them even if they twisted away, blinking at him with sleep-red eyes and tight-shut mouths.

And she would have a sudden memory, of when she first came to live with him herself, and a complicated anger. She should be happy he was so solicitous, so tender toward his children. So many men were not. But what did he think, that she was like a stepmother, uncaring? That she had not fed them?

A terrible pregnancy.

She could not face food, so every day she walked out into the garden and fed her share to their new calf.

A terrible pregnancy, but a good birth.

She called the girl Zenna-Mariam. News of Mary.

On market mornings her house would fill with running feet as the bigger children leapt from their beds and raced past the smaller ones to take up their places on the outside wall. She did not follow them, but if she stood at the window she could see into the market too, see the egg-sellers and drovers, the chilli and butter merchants, the peasant women guarding little heaps of spices. The morning sun shone off the high pale minaret of the new mosque – built on land given to the Muslims of Gondar by the Italians – and set aflame the perfect dome beneath it. It sparked off the buttons and belt buckles of the soldiers, and of the officers from the prison, which had been moved to the former Italian consulate just around the corner. And it lit the municipal gallows, and the men these soldiers and officers were bringing to the gallows.

Sometimes she watched, unable to look away as their bowels failed or they cried out blessings and imprecations, confessions and protestations, cried out to the skies or to their milling half-interested audiences, I found him with my wife, he deserved to die, no blood will spill, only milk, I am innocent, I am innocent, as God is my witness I am innocent. Sometimes she too cried out, involuntarily, Oh, think of your mother, think of your mother, where is she now? But mostly she would throw herself onto the bed and cover her eyes and stop up her ears and pray until that week's executions had ended.

Even before the Italians had finally surrendered, the emperor, knowing that his absence meant loyalties were even more fractured and contingent than they had been before, prosecuted with zeal the process of centralisation he had begun before the war. Menelik II, Empress Zewditu's father, had been the first sovereign to require taxes to be paid to the state as well as to regional lords, but some sacks of grain and hours of labour were nothing compared to what Hailè Selassie, after a short initial period of caution, now required, and in cash rather than kind, cash which no longer flowed through church and lords, but directly to Addis Ababa. Diligently he drew to himself economic control; assiduously he tightened his command over individuals, reinstating powerful nobles even if they had fled when most needed; sending trusted allies into the regions; promoting those who had abandoned him or been tempted to abandon him, bridling them with their own guilt and holding them close; playing, with a shifting mixture of flattery, reward and force, on the vanities of the men, often humbly born, who had become (often well-loved, well-respected) leaders in his absence.

When Emperor Teklè-Haimanot built Ba'ata and settled upon it the rich lands of Bisnit, of Dembiya and Deresgé, he had also given its aleqa the job of judge in the Saturday market, with the right to claim a third of all tax income from goods sold there. Under the Italians such taxes had ceased, and then her husband had, of course, lost his aleqa-ship. Within a year, however, the emperor, recognising that her husband had, during his decades of training, learned all the books of the Fit'ha Negest by heart, appointed him a judge in the provincial criminal court.

Now every Saturday after breakfast her husband took his seat beneath the huge sycamore fig on the western edge of the market, and litigants came to stand before him, shoulders thrown back, right feet forward, grasping the tied-together tips of their shemmas, hands trembling with emphasis. 'May God show you!' – that the cow was really theirs, or the money; that it was they who had received the criminal insult, not given it. That the land was theirs, and the tributes from it – in Gondar the emperor's new ideas were still often treated as inconvenient distractions from the main business of enforcing rights that had existed for centuries. Those particularly convinced of their position might throw their shemmas off their shoulders as if readying themselves for a flogging. See? Look what I am prepared to suffer. And Aleqa Tsega would pass his judgements, imposing fines, jail terms, taking as payment whatever the litigants could offer: teff, a sheep, a couple of chickens, a promise of wheat at harvest time, or a handful of well-worn silver thalers.

Often he took Edemariam along, to help, and watch, and learn. Edemariam was nine years old now, with a dark thin face and a solemn gaze. He had begun his alphabet five years earlier, sitting cross-legged with the other four-year-olds under a tree at Ba'ata and singing the letters in a ragged echo-chorus. Ha-hu. *Ha-hu.* Lè Lu. *Lè Lu*, while the priest-teacher paced before them, stick at the ready. Ha-hu. *Ha-hu.* Lè Lu. *Lè Lu*. The alphabet, the Apostles Alphabet, the first epistle general of St John. The psalms of David, written on goatskin, wrapped in a leather carrying case and hung around his neck – when he learned to sing these, she had cooked him a small feast in celebration.

They had hired a tutor who came to their home, a man with a pronounced limp who arrived each morning to lead Edemariam through the chants composed in the sixth century by Yared, a man so holy and with a voice so luminous it was said his king, listening, had leant on his spear and pierced his foot and neither man had noticed. She kept an eye on her son as she went about the household chores, and saw a healthy boy with a sensitive side who, especially when he began to join a few other children at the teacher's small house in the grounds of Abiyè Egzi church, did not hide that he was afraid. It was the dead people in the cemetery, he told her, weeping. He expected them to rise out of the ground to chase him. She wasn't surprised when one day he refused to go at all. And because she believed it was not helpful to gainsay a father's discipline she tried not to show the empathy that rang through her when her husband found out and dragged Edemariam, over and over again, through a bed of nettles. But it was cruel. Really it was cruel. (A thought that did not stop her when one day some years later she found Edemariam had stolen a coin and used it to shoot air rifles instead of going to school. She had dragged him home, whipped him, and tied him to a chair in a storeroom before presenting him with the family confessor, who extracted a promise that it would never happen again.)

She tried to fix her mind, instead, on the intense pride with which her husband watched Edemariam, sitting beside him in the market, fill out receipts and copy court documents. When he came home full of his pleasure, however, she listened and frowned, and said, you must not tell him. He will become conceited; love him silently, in your heart. She did not tell him that her own heart had brimmed with a delight that echoed his.

But even Edemariam's ready help – and, eventually, a trip to their confessor to get her father-in-law's curse exorcised, so her husband could finally write – was not enough to get through all the work, because while the mayhem after the war had changed character, it showed little sign of abating. True, they had a new governor; the emperor had appointed Ras Kassa's only surviving son. Asratè, however, who had spent the war in England and Jerusalem, was very young, and his concern was not always with a city whose roads were potholed by tanks and the floods of the rainy season; whose bridges had been blown up by retreating Italians; whose banks did not function; whose municipal buildings were being used as byres and lavatories; and whose hospitals, already stripped by the British (arguing Africans had no mechanical aptitude, they had taken equipment and medicines, along with most of the contents of Italian-built factories, to their colonies), were now required to send many of their remaining beds to Addis Ababa. Asratè was proud of his lineage, well aware of his family's claim to the throne and of the power it thus held in the country and, however subtly, potentially over the emperor. And unlike his father, he was inclined to test it: he was rumoured to have joined in the general seizure of corrugated iron, and when the crown prince sent 10,000 thalers with which to repair Gondar's roads Asratè appropriated the money for himself.

Within a couple of years Asratè was replaced by Ras Imru, the emperor's second cousin. Aleqa Tsega bought an ox; she supervised its slaughter and an arrangement of choice cuts to send to the palace in welcome. They knew Imru and the emperor had been raised together, that both were described as modernisers, that in his postings before the war Imru had attempted to institute fairer taxation and encourage modern education, with some success – except in Gojjam, where he was resented as an outsider, and his attempts to rule by consensus disregarded until he resorted to force. Gondar, which refused to forget that it had run the empire for hundreds of years, was proving equally insular and reactionary. Imru's census was sabotaged, his proposals for reform fell on politely deaf ears.

No one could ignore, however, his method of dealing with looters and brigands, the way he took on the slave traders.

The bodies were often left hanging in the market for days.

Even those who had not known that the emperor was due to visit Gondar were left in no doubt when, five years after the Italians fell, certain roads suddenly improved beyond all recognition. Tarmac swept from the Piassa down to the Saturday market, then curled back up to the castle compound and into town. Side roads were swept, trees planted and watered, and her husband and his colleagues composed wide welcome banners in Ge'ez to hang at the entrance to the castles, across the municipality buildings, across Samuna Ber, the Gate of Soap at the south-west corner of the city, and over the entrance to the new airport building whose delayed opening was ostensibly one reason the emperor had not come until now.

'Emperor Fasil took a year to reach Gondar,' proclaimed her husband, at the airport to welcome his liege, yet again at the head of his priests. 'He stayed three months at Yibaba, three months at Irengo, three months at Denqez, and then finally three months in Gondar, but your majesty has been seen in both Addis Ababa and in Gondar within the hour! The king of kings has pierced the clouds and arrived among us, and our joy has no limit. Long live Emperor Hailè Selassie! May the kingdom of Ethiopia endure forever!'

The students clapped and sang, the women trilled their ililta, the patriots discharged their rifles into the air, and the emperor and his empress processed into town, where they feasted Gondar's notables with all the generosity required of them.

She was not there, of course, but she paid close attention to the imperial movements. She knew that sometime in the quieter days of the two-week visit, the empress, who like her husband had to be seen to be pious, would come to Ba'ata to pay her respects, and she wanted her daughter to be ready.

Alemitu was nearly sixteen, a tall fierce girl beginning to possess a commanding beauty. She was beautiful and she was bright, and Yetemegnu had taught her to sew and spin and embroider, to weave baskets tight and gleaming. She had tried to protect her, refusing to let her play with other children because she saw how they bullied her and called her names she couldn't hear, dragging her away when she approached the fire. 'You're already deaf! You can't burn yourself as well!' And then she heard the empress had established a school for handicrafts in Addis Ababa. She could not think how to enrol Alemitu – until the imperial visit was announced, when she went to their confessor. Tell the empress my daughter is deaf, she said to him. Ask her to take Alemitu to Addis and educate her. And he had promised, yes, of course I will do that.

Now she would have to do something she had never done before, loiter with all the other petitioners in the churchyard for a scrap of the empress's attention, but she had confidence that at least the ground had been prepared. Carefully she and Alemitu arrayed themselves in the whitest of white dresses, the best necklaces, the finest shawls. Then they walked to Ba'ata and found a spot under the trees where they could wait.

They watched as the empress arrived and Alemitu's father stood in the knot of priests and nuns that came forward to greet the empress's retinue, watched him stand silent as the new aleqa, and then a priest, interrupting the aleqa, and then her husband, talked and bowed and talked. The empress stood listening. A pale woman, big, with wide shoulders and hooded eyes. Full lips, a short neck, and double chin. Not handsome, but a strong presence. When the men were finally silent, their empress walked up to the unfinished church, bowed her head, and kissed its walls.

When at last the empress moved toward the gate Yetemegnu pressed forward and bowed. She is deaf, madam, she said, pushing the teenager in front of her. It is this child who is deaf. The empress looked straight at Alemitu. Yetemegnu was so pleased and relieved that for a moment she did not notice the woman was already turning away, that an attendant was holding out a hand, and that in the hand was money. When at last she bowed her head and put her own hands together to receive it she was staring at a ground she could not see, for rage, and for humiliation.

She was becoming accustomed to a bit more freedom. Partly it was that her husband was again often away. Sometimes their son went with him, describing on his return rowing across Lake Tana in low reed boats and scrambling ashore to talk to the abbots of quiet churches and quieter monasteries. Or her husband went on long trips into the interior to do she knew not what, only that sometimes he took with him large amounts of money, which she assumed were from the emperor. There were mornings that began with clopping and whinnying and milling priests, and one man carefully balancing a tabot on his head; then everyone knew Aleqa Tsega was travelling to the Qimants and the Jews, to summon them down to a river with the blare of a trumpet and baptise them. She had even heard one story of so many coming to take their first communion – because otherwise they would not be allowed to own land – that her husband had gathered them into groups and given all in each group the same christening name.

He made frequent journeys to Gonderoch Mariam and to Addis Ababa, where he haunted the church headquarters and palaces and spent hours talking church politics and poetry with his childhood friend the dean, who had just been promoted to bishop of Harar and renamed Theophilos. She did not know that at this time her husband proposed the idea (controversial generally, but in certain quarters of Gondar anathema) that church tithe income should be converted to cash and paid as salary to church employees rather than to anyone who might lay claim by reason of birth. She did know, however – if only because he glowed with righteous vindication – that he had asked for the return of Ba'ata, and the request had been granted.

His jealousy, if not exactly easing, was, after the moment when she had faced him down, no longer expressed in blows. She could even have her own visitors now, and she came to love the point at mid-morning when she could turn away from chores, take a handful of green coffee beans, release them rattling onto a flat metal pan, and settle it onto the low fire. Gradually the smell of roasting coffee would lift and weave through the house, a pleasure heralding pleasure, cut through with the scent of fresh grass and wildflowers, the outdoors brought inside and strewn across the floors. When the visitors arrived a servant would carry the pan in for a moment and with a cupped palm direct the smoke toward each person, so that they could smell it. Then the beans were taken away and a muffled pounding underlined the preliminaries of conversation: Are you well? I am well, thanks be to God. And your husband? Very well, may His honour increase. Are the children well? All well, all well, may His name be praised.

For a long time she just listened, pouring coffee, or, growing into her role as host, nudging the conversation along – And what happened then? Oh, may the Lord keep them safe! – watching faces, listening to the sometimes dissonant interplay of voice and expression and story, hoarding everything away. Who had fallen out with whom, who was related to such and such; who was ill and who looked unfeasibly happy; who had been seen at church and who had not. And did you hear what she said to him? It seemed so innocent, and yet he caught the hidden meaning, and it drew such blood. And she would laugh, released into a kind of pure appreciation.

As the second and third cups were passed around she sometimes joined in the conversation herself – tentative at first, but with a growing realisation that she was mistress here, and they duty-bound to listen. And as she became surer of her stage, she began to feel her father's skills surfacing – to discover that language came easy to her; that she could spin out a tale, and, drawing her listeners along with her, wrap their companionship about her for that much longer.

But she also began to learn she had to be wary: that her position, and particularly her husband's, meant those she thought were just neighbours, thoughtful and generous, often wanted something: preferment, connections, favourable judgements. Sometimes it was easy to tell. When donkeys appeared at the entrance, for instance, carrying firewood, honey, butter, the softest, whitest cotton blankets, quintals of raw green coffee, or a sheep she did not expect was dragged bleating through the gate.

She had seen how her husband reacted to this largesse. She had tried to intervene as he shouted at a cowering monk, threatening to throw his hundred paper notes – enough for more than thirty fat sheep – onto the fire because he had had the temerity to attempt to bribe his way into a job as a vicar. A murderer's brother had his hundred notes flung into the air. She had joined the entire household in dropping to her hands and knees and picking them up for him before he scuttled out.

So whatever came to her she sent smartly back. But sometimes she felt irritated. Everyone else took these inducements; they were so common in the courts they were almost considered fees – why did he have to be the only one who refused? Once a priest she knew slightly came and sat in her reception room and fished out from among his robes a thaler's worth of the best coffee. 'I know you love this; it's fresh from Jimma. Let's drink it together when I come to visit.' She refused, and he left, but came often after that, to sit, and chat, and eventually – what harm could it do? – she accepted enough beans to make a glorious pot of coffee.

What does the priest want? she asked her husband one day. Can't you finish your business with him, and send him away? He looked at her. Why do you care? You don't normally ask these things. I'm just curious. Did you take a bribe? Swear you didn't. But she could not swear, and finally, fearfully, she answered, he begged me to take coffee with him. I will never do it again.

Her husband laughed. But this is how bribes work – don't you see? They inveigle their way into your kindness – never do it again. And again, she promised.

In the afternoon she sat in the shade in the back yard. The sky was high and deep and blue, and the hills were parched. The birds were silent in the junipers and a breeze tickled the leaves of the young eucalyptus trees, making them shiver in the sunlight. They were coming on well – it had not been long since they planted them, he scooping a hole out of the ground, she following with the seeds, placing them exactly, covering them over. They would never need to buy wood now, for fire or for scaffolding.

The basket of barley on her lap had a comforting heft. She combed through the grain with her fingers, picking out crumbs of red earth, errant stalks, husks dry and light as ladybird wings. Every so often she took a few handfuls and spilled them into a smaller, flat basket, then shook it from side to side until the barley lay a kernel deep. A snap of her wrists and the barley described a parabola, twisting through the afternoon, perfect but for the faint fuzz of dust sloughing off into the air.

She soaked the kernels overnight, drained them, shut them up tight in a covered barrel. After three days they began to germinate, white roots clawing out and tangling with their neighbours'. She enlisted the help of anyone available, and together they lifted up the mass of growing matter and spread it out on wide false banana leaves on the ground. She covered it over with more leaves, and then with stones, flattening the sprouts into a wide malt cake.

There was so much that when she reached up to store it in the kitchen rafters for the rainy season, it formed a second roof.

She said to him, in passing almost, because she thought she knew the answer, my brothers are marrying, two of them. I would like to be there to see. You must go, he answered. Take Alemitu with you. He borrowed a mule for her, and she and her daughter rode through ripe barley fields to Atakilt Giorgis, where two oxen had been slaughtered, and a sheep, and the barley beer and honey wine flowed for four days. On the third day, she stood and made her way toward a circle of clapping, dancing women. For a long moment she watched, then moved forward, and the women, feeling her approach, parted to make a space for her. Hands on hips. Shoulders down. Body steady. Pause. Jab, with her chin, forward. Jab, to the side. Forward. The other side. Faster, faster, the rest of her body motionless. The circle reformed, recognising her gift and placing her at its centre. She danced until her face broke open with laughing. She danced until she could barely move. A brilliant wedding, everyone said afterwards. Such a brilliant wedding. A brilliant wedding, she replied.

When she returned she began to test, little by little, the circumference of her liberties. She shopped in the market. She redraped her shemma so it covered her shoulders and her hair, and she went to Ba'ata, bowing at the entrance, kissing the outer walls, walking home again. She took communion, and later she took her children to communion also, tying the smallest onto her back with a broad shawl, letting the others follow her through the streets at their various speeds.

She began to go more often, each Sunday if she could, and if he was away, during the week as well. She craved the moment when she could step through the rough gate and into the church compound. Birds chirred, roosters crowed, and if there was a service in progress, a drum thumped the de-dum de-dum of a heartbeat from the holy of holies. In the wet season the rain dripped from the jacarandas and the faithful shivered under the eaves; in the dry season they searched out the shade. Outside the bethlehem nuns picked through wheat, and the occasional gelada baboon bounded through the tops of the fern pines.

To the south doorway: lips and brow, lips and brow, against the rough stone, then, discarding her shoes, into the sweet-smelling dusk of the church. As her eyes adjusted she would turn toward her favourite spot, under Mary robed in blue and red and gold, eyes large, face pale and shadowless, holding her grown son. The

Virgin was guarded by angels and surrounded by more frescoes which, because Yetemegnu could not read, were in effect her Bible. She would sink onto her knees. Her mind, busy with chores and children, would begin to follow the chants, the drum, and to still. Her stomach would growl, and she would feel a kind of satisfaction at this reminder she had fasted the requisite eighteen hours, had asked forgiveness of anyone she thought she might have wronged, had kept herself away from her husband, had worn her very cleanest clothes. A kind of peace crept through her.

The sun would be high by the time she left the church. Everyone she passed bowed, How are you? Are you well? Your husband? Your children? Your cattle? All well? and she would bow back, murmuring, Well, thanks be to God. All well. And you, are you well? When in the final decades of her life she found herself living far from home this was the thing she missed most: the sense of belonging recognition bestows, and her spot near the south door.

The heavens no longer cracked open with thunder that made her bury her head in her shawl and call on all the saints for mercy. The last hailstorm shredded the collards and left hard white drifts under the eucalyptus trees, but that was a few weeks ago. Now rainbows vaulted through the mountains and she taught her children to sing, 'Mary's sash is for me, for me, Mary's sash is for me.' Bullroarers echoed across the fields and from behind the house came the arrhythmic chop of wood being split.

Her storerooms filled. Broad beans and peas from the market. Lentils, chickpeas, the first barley harvest. Garlic, shallots, ginger. Bags of long hot chilli peppers. It took a full day just to cut them up and spread them out in the sun to dry. The yard began to disappear under squares of green and yellow and red.

Then it was the turn of the spices. Dirty gold rhomboids of fenugreek, dry-roasted to deepen colour and flavour. Bishop's weed, unassuming yet pungent; striated spheres of coriander seed. Angular black cumin, love-in-a-mist. Basil both sweet and sacred. Beautiful mustard seed, tiny rolling orbs of grey and umber and brown. Rue seed from the bushes that grew about her front door. False cardamom splitting under her hands to reveal dark kernels huddled into pale white nests.

When the peppers had dried to a deep blood-red and been pounded into flakes, when the chickpeas had been husked and dried and split, she began to compose the things that would inform much of what she would cook: berberé, aromatic chilli powder; and shirro, spiced chickpea flour. Ample amounts had to be provided, of course, and more than ample, but she knew that along with volume, the subtle tastes of these things in particular would be a way for hundreds of people, the most powerful in the province among them, to judge both her and her husband, the richness and status of their house. She knew that because her husband was not from Gondar it was not enough to be equal, that their hospitality must be outstanding. But she also knew she was now good at these particular arts, and that she had come to enjoy the hours she spent adding a touch of cumin here, a handful of crushed ginger there, a pinch of salt or a dusting of rue. Finally the berberé and shirro went out into the sun again to dry.

And then the sound of grinding, of stone against spice against stone, filled the days.

There were still unexpected guests at mealtimes. She had kept the habit of watching out for her husband from the window, so by the time he and his guests had crossed the threshold the low stools would be arranged around the mesob and the sauce she had made that morning would be bubbling up. Servants would appear, bow, pour water for the washing of hands and dusty feet; the injera would arrive, the slim-necked flasks of mead.

She would ensure everyone was served, making the ritualised fuss, working through the catechism of offer and refusal – eat, you must eat, upon my death you must eat – that in this region preceded any breaking of bread, regardless of how hungry anyone might be, until they finally acceded that yes, they might taste just a little, just a tiny bit, just for her. She would hover, provide the generous portions that would in fact be consumed, accept a couple of mouthfuls put together for her by her husband, but she would never sit down.

As their stomachs filled and the mead warmed their veins, official business would shade into gossip. Who was looking for promotion and hadn't a chance. Who had been seen eyeing up whose wife. Who had come out with a surprisingly sophisticated qiné. Those from Ba'ata, especially, returning with pride to their alumnus Aleqa Gebrè-hanna, whose puns and jokes had attained the status of folklore and were known by every schoolchild in the land. Did you hear the one about when he came back to Gondar, and ran out of money? He was out of favour with the emperor at the time so he sent his wife to the capital to tell Menelik he had died. Menelik, feeling guilty, sent her back with a purse full of thalers for the memorial service, and was understandably annoyed when, some months later, Aleqa Gebrè-hanna reappeared in Addis, alive and well. 'Your majesty, they had so many rules up there,' said Aleqa Gebrè-hanna, pointing to the heavens, 'I preferred to return and live under yours.' Billows of laughter would break through the house.

Then again, said one – a glance askance at her husband – our current aleqa isn't too shabby either. Remember when he met that other aleqa in the roadway, who asked, 'What are you doing, so far out of your way, and so late into the evening?' She tensed. No one but evil spirits roamed at night, and Gojjamés were often accused of being evil spirits; the slur on her husband was clear to everyone. 'And Aleqa Tsega replied, "Looking for a donkey"' – for evil spirits attacked those who resembled donkeys. She watched as they laughed, too uproariously, flattering their host and his sharp wit, seeming to give him the spoils of victory. She saw how her husband's answering smile did not touch his eyes.

And she remembered the old Gojjamé woman who had lived in a little room next to their compound. Her children were grown and gone, and she was destitute, so they had not been charging her rent. When she died Aleqa Tsega had wanted to give his country-woman a respectful farewell, so they had laid down carpets from the church on the floor, beautiful Turkish carpets, and asked all the neighbours to the wake, and the churchmen for absolution. But because she was poor, and from Gojjam, no one had come.

Her husband had surveyed the empty room and said, Will this happen when I die? She had tried to reassure him, but she too had been shocked by the nakedness of the rejection, and knew she did not convince. She was not surprised when, a short while later, he began to have regular meetings with three other Gojjamé men, in which they decided to pool practical resources, to exchange infor-mation, to donate money to a fund that would provide loans to any Gojjamé person in adversity, and would be held in trust for their funerals.

Now the young boy cousins began to arrive, and the farmer relatives, walking in from Dembiya, from Gonderoch Mariam, from the
archipelagos of hamlets in between, each with a staff slung across
his shoulders, and on the staffs heavy hide sacks. When she
reached into one and drew out a piece of honeycomb she saw
some bees had not escaped the smoke and were curled, still, in
their cells. She felt sorry for them, deprived of their holy handiwork
and of their families, and hoped they'd had a quick death. With
a finger she caught the dripping honey and brought it up to her
mouth. It tasted of sunshine, of breezes scudding across mountain
meadows.

The buckthorn was nearly ready. The bushes in their compound
had been stripped of their leaves, and more had been bought in
the market. For a few days now they had been spread out in the
yard, drying.

Weeks ago she had sowed a patch behind the house with
pumpkin seeds, then had gone out each morning to check on
seedlings, to lift widening leaves and peer at the fruits growing
underneath. She would bid them good morning, ask how they
spent the night, had seen so many develop that standing among
the searching tendrils, counting them, she had laughed to herself
in amazed delight. But now they had been harvested, the soft
watery flesh cut into cubes that lay in the sun, shrinking and
sweetening.

When Alemitu turned seventeen Aleqa Tsega chose a student of
his, a clever one who could read and write, and told his eldest
daughter this would be her future.

One of Yetemegnu's half-brother's daughters was marrying too,
so they threw a huge party. The food was rich, abundant – whole

sides of oxen, perfectly seasoned chicken sauces, all the parts of many sheep, each dish ringing all the available notes of flavour and texture and skill. The das stayed up for days, the music rarely ceased, the women danced. The men took it in turns to declaim couplets they composed on the spot, about the mead, the food, the bride. Yetemegnu stood, and went to join the dancers, the half-sisters, cousins, nieces, the neighbours and the neighbours' daughters, and they welcomed her in.

It was a slackening in their attention she felt first, a stumble in their support, and a dropping of their gaze. She turned to her husband. His face was closed tight, his eyes small. Anger flared in her at this again. I thought we were finished with this. Look, our first daughter is married, silent as custom requires, there under her muslins. Surely it is right, necessary even, that I should take this lead?

Then she heard the laughter. She looked around. One of the men, a relative from Infiraz, stared back bold and straight. Pleased with himself, he repeated his verse:

> You have a horse and a mule stabled in your byre.
> Why go looking for any other animal?

There were gans everywhere. Stacked up under trees, ranged along the outside wall of the kitchen, a squat pottery army, awaiting orders. All day now a fire burned, and one by one the gans, their insides washed in ironweed and soapwort, were rolled over and suspended above the fire. They swayed, gently, drinking up borage smoke and sweet wild olive.

The storerooms were overflowing. Green and black chickpeas, white broad beans, hard yellow split peas, maize. Wheat and barley and sorghum. Finger millet. Shallots and garlic and ginger and coffee. Limes she picked up and held to her nose, taking breaths so deep they made her dizzy: the advent fast had begun, and no one ate till past midday. She tracked household expenses only, but she knew these intoxicating smells and colours represented a large proportion of a year's income, much, in fact, of what wasn't being spent on his church. The responsibility tightened her shoulders, hurried her steps, began to enter her dreams.

She ground the buckthorn to powder, mixed it in water. She soaked the maize so it would sprout, and the millet. Then she turned her attention to the mead. This was the more precious drink, intended, until recently, only for aristocracy, but it was simpler to make: just water and honey for the moment, three parts water to one part honey, stirred and covered and left to ferment. She took extra care these days though – she remembered how, a couple of years ago, Ras Ayalew had called her husband over in the middle of the party and pointed out that the beer was mellower than the mead. Her husband had bowed and smiled and made admiring comments about the ras's refined palate, but later had shouted at her in embarrassment and demanded it never happen again.

When three or four days were up – she could tell, from the smell, when it was time – she went looking for a relative tall enough to reach the hardened cakes of barley sprouts, now perfectly smoke-touched, down from the rafters. She pounded some into pieces, then added them to the beer barrels along with a crumbled pancake of wheat and finger millet, the germinated (and now dried and pounded) maize, and a little more ground buckthorn, sprinkled on top and swirled with a stick into the darkening mixture. A handful of fine-ground buckthorn went in with the honey, too, and more water, and both were covered again. Within a few days the smells of beer and mead rose, mingling and drifting throughout the rooms.

In her ninth pregnancy she spent much of her time spinning. She was a grand lady now: her husband was comfortably established, or seemed to be, she was in her early thirties, she had servants and growing children who helped with the work – those who were not in school, at least. Her husband, seeing on his trips to the capital how the emperor favoured modern schooling, how increasingly he sent the brightest abroad to study, then brought them back and appointed them to the highest posts, had taken his sons out of the churchyards and enrolled them in modern elementary school. So now she could more often do what was expected of women of her station: sit on a daybed, and direct operations. From the basket at her side she took a cloud of cotton, thinned where the dark seed had been picked out, and snared it with the hook on her spinning reed. A quick twist of her right hand, and the cloud shrank toward a point. Her left hand steadied the cloud, let it out gently. Another twist, and another, until her right arm was stretched out straight and between her hands quivered a white thread, tiny fibres along its length escaping and catching the light. She wound it tight around the top of the reed, picked up another cloud of cotton, and began again. Eventually there would be enough for a winter blanket.

One afternoon she was sitting as usual on the daybed when she heard someone calling, from outside the house. The voice was urgent, frightened, but also somehow thrilled. She pulled herself upright and tugged at her dress to right it. Because the feast of St John was approaching it was a new dress, embroidered down the front and at the hem, which rose in a curve that echoed her growing belly. She made her way outside, to one of the shacks in the compound they often lent to visiting relatives or to acquaintances in need. Except for pinpoints of sunlight that had forced their way around the nail-holes in the tin roof and walls, the shack was dark when she went in, pushing past holiday torches of bound sticks that rested against the doorway, rushes tied with yellow masqal flowers. When her eyes had adjusted she saw the still shape in the bed, a length of greater darkness solid in the gloom. She bent over her half-sister and listened for breath.

She dropped to her knees. Bring coals, incense! Perhaps they would help to revive her. Duly they were brought, and she set about organising them, praying aloud. Oh Mary mother of light, come to us now. Queen of Heaven, help us. The lit coals revealed ritual detritus: a coffee tray, cups smeared with dregs, grass on the floor, a dropped necklace of coloured beads. The incense rose, sweet and strong. She felt herself swaying; her own voice, praying, seemed suddenly far away. Pain stabbed through her flanks.

When eventually she woke she woke slowly. She swallowed, and her throat was dry. Her knees throbbed. Her stomach growled. Her neck ached, and her shoulders. She felt incredibly tired; depleted, scooped out. But light, too, as if she had been dragging a heavy load up a mountain and it had been lifted from her, so now she was standing in her shift in the high bright air.

She became aware of someone sitting in the corner, and lifted her eyes to her husband's. Worry, and a deep irritation, stared back at her. But something else looked out at her too, something she had not seen before. Was it fear? Respect? She registered it and stored it away, taking it out every so often over the weeks and months that followed, turning it over and over, holding it up to the light.

Finally cooking began. The first dish was shirro mixed with berberé, onions, niger oil and a little water — no butter, because the actual festival always fell in fasting season. She stirred it on a flat clay pan over the fire, stirred until the shirro was absorbed, added more. Stirred until it was absorbed, added more. Again and again, until it sat in soft, crumbling piles.

She measured out flours, barley, wheat, broad beans, and beat them with oil and water. Siljo could not be granular, it could not be watery; it had to be delicately spiced with red onion, garlic, ginger; pungent cabbage seed, coriander, rue; then berberé and sweet basil and crushed mustard seed, added right at the end. By the time it was poured into a madiga and sealed for six days, her waist hurt. Her arms hurt. It hurt to lift her children.

When he went out each morning he would say, Make coffee. And she would do so and wait for him. She would make the breakfast he loved, too: torn-up pancake, thin, almost as thin as injera, softened with butter and spices, making it as tasty as she knew how. And then he would take hours to come and she would swear under her breath, God roast you, is this how you treat me?

When at last he arrived he gave no explanation, and she said nothing either, but reheated his breakfast and placed it before him with a slight bang.

It was weeks before she discovered what he had been doing, when she walked with her servant to a communal well to get extra water and said to the other women, surprised, Where's all this from? Who's been building rooms on our land? And they said, 'Why, they're yours – didn't you know?'

So that was what you were doing while I was sitting at home cursing you, she said to him. Why didn't you tell me? What are they for? They're for you, he answered. You can rent them out for income, when I am gone.

A girl was born to them, keen-eyed, good-tempered. Yetemegnu called her Maré, my honey. And a boy to Alemitu. Her first grandchild.

Edemariam graduated from grade eight and boarded an aeroplane for Addis Ababa because there was no school in Gondar that could educate him further. She dressed him in a new wool suit; she slaughtered a sheep and fed him his favourite dishes. The entire neighbourhood came to see him off.

Outside they were building the das. The skeleton was nearly assembled – tough slim poles of eucalyptus cut from their own trees and lashed together with fresh eucalyptus bark: uprights, horizontals, beams waiting for their blanket of branches and cured hides. Rattling leaves covered the ground; anyone who entered could not help but crush them underfoot. The earthy, clarifying smell was everywhere.

The mead was ready. She placed a piece of fresh muslin over the mouth of the gan, but it took several people to tilt the huge vessel and pour it into demijohns. When they were finished the glass containers sat in amicable rows, their bases smooth globes charged with liquid gold.

Now whenever she began to yawn, great yawns that took over her face and stretched her shoulders tight, when he saw that though she was there in body everything else was withdrawing, curling away, out of his reach, he would say, spread grass for her, give her incense. Then he would pick up his gabi and walk out the door, into the drawing dusk.

The younger children would huddle at the edges of the room, wary and fidgeting. Sometimes there were neighbours too. They would watch as she set out the coffee tray, counting out the little cups, arranging them just so, pausing to yawn, again, and again. Incense coiled through the roof beams and coffee beans popped on the fire. She knew the children were observing every move, registered one or two beginning to whimper, but exhaustion was stealing through each sinew. It dragged a mist through her mind and held her down.

Down, down, down.

Take your shawl. Cover your head, cover your face. Sink.

Her neighbours told her what would happen next. She would climb off the daybed and onto the earthen floor and then, slowly, would begin to move, almost imperceptibly at first, then faster, shoulders revolving about the fixed point of her waist, faster and faster until she tipped forward onto her knees. Faster and faster and faster, until her torso scrawled ellipses through the thickened air, head following a half-beat behind, back arching like a horse ridden hard across a plain, a horse responding to the sting of a whip on his hindquarters, curving sharply in then out again, galloping on and on and on.

And then the sounds would begin, cries harsh and urgent, turning to fierce song deep in her throat, songs like nothing in the daylight world, or even in the dark mead huts or the smoky churches. Songs and poems full of wilderness and nights in the open, of words that promised familiarity, but slipped out of comprehension. Songs tender and loving that curled around the children and brought their mother close, to touch their shoulders, to kiss and bless them. Songs shredded by declamations, a staccato of demands and sometimes of prophecy.

Once her husband had been present when the zar took her, as it had begun to do more and more frequently since the day she had found her half-sister possessed and, in running to help her, become infected herself. He had watched until he was on his knees too, on his black cape spread out across the floor, begging God and all the saints to free her from this satanic thing, for who but the devil could play upon a woman in this way?

But the spells continued, and he acceded to the zar's requests, buying her fine shawls, a necklace of black silk wound with gold filigree, the finest coffee. He could not but notice she was beautiful after the spirit had taken her, that a private calm brightened her face, but noticing only deepened his distress and his dislike. Here he was, a priest and scholar, a man wanting to bring modern ways to a hidebound clergy, and here was his wife, dancing with pagan spirits, speaking their tongues, ecstatic in places into which he could not follow. And he would kneel in his church, praying and weeping, or beg their confessor to help.

She took holy water as the confessor prescribed, asked Mary for help again and again, but still the zar came, and took her away from him, shaking her slight body, calling through her voice, leaving her limp and exhausted the next morning, unable to remember what she had said or what she had done but knowing, even if she could not say it to herself, that what she felt was a kind of release.

*

She looked across at her husband. Under the untidy turban shadows deepened his eye sockets and lengthened his nose.

The lamp guttered. His turban climbed up the wall, then dropped. When he took another page from the sheaf of papers before him she saw his hand was shaking. He said nothing. Had said nothing, for hours, just leafed again and again through those papers. Outside in the star-hung dark a dog howled.

Let's sleep, she said. It's nearly eleven, and we should get some rest. Let's douse the lamp and sleep.

It took a moment to feel the sting of the slap, she was so shocked, and then another moment to react.

She sprang to her feet. Really? *Really?* And still he said nothing.

The fishermen were a procession of walking scales: long rods balanced on shoulders, at each bobbing end a full basket. A mass of eyes and tails, of silver and grey and dots and stripes. The smell of wild fresh water.

She divided the fish into three piles and started washing their slippery bodies, splitting them open, dragging out their guts, cutting the flesh up small, removing every bone. It was close work, messy. Her arms flashed with scales and her fingers wrinkled in the constant wet.

Around her the noise levels rose. Almost every woman she knew was here to help. Neighbours, relatives from the country, their neighbours, their servants. They came with supplies on their heads, children on their backs and at their skirts, and divided the cooking between them. Peeling, pounding, sifting, chopping. Especially chopping. Quintals and quintals of shallots, ginger and garlic.

It took her a while to sort the fish, because she kept being interrupted. Where was the salt? How spicy did she want the pumpkin? Could she taste this? Did it lack anything? She gave each query her full attention, thanked each messenger graciously, intervened without compunction if she felt something was going awry, dispatched servants hither and yon.

When they first began having these yearly parties relatives had taken over, bossing her about as they would any other child. But as she grew into her role, and especially since her husband was re-awarded Ba'ata, they had retreated. Now it was she and only she who directed operations, who knew herself to be in charge even when all she was doing was chopping garlic and listening, to the chat, the gossip, the songs. Would that every day could contain such camaraderie, such company! And, if she was honest, such a public stage for her skills and her position.

By the time she moved on to the last pile of fish, to dip them in seasoned flour for frying, the sun had set. The women had decamped into the das. There was less talk now. Oil lamps shhhhed into the silences and cast giants on the walls. The corners filled with dark.

It was near midnight when they finished, and the food was set out in row upon row of clay pots and baskets and trays and bowls. Those women who could not go home found beds in the main house or sank down where they were, pulling white shawls over their heads. She waited until they were asleep, stirring this, tasting that, covering everything to protect it from flies. But finally she stopped, and took one last look around.

Then she blew out the lamps.

They are all risen against me, he explained, weeks later. I have already protested my innocence before a judge, who believed me, that's what the papers were, but they are coming for me again. What shall I do?

It was she who was silent this time, though the tears flowed down her cheeks.

Come with me, she said at last. Come with me to Ba'ata.

They made their way to the far wall and roused the hermit who lived there, in a lean-to of rocks and branches. She had been here before, when the zar first took her, had stood before him, knees knocking, saying, this thing is persecuting me. She had stared at the ragged animal skins he had jabbed into a cape with conker-berry thorns, his dirt-dulled, matted hair. Eventually, when she had begun to think there would be no answer, he had spoken:

Wild animals were chasing me, harrying me,
tormenting me,
Yet the innocence of the dove brought her to my side

There was nothing more, but she had returned home wrapped in fragile hope, she could not have said for what.

So now she returned with her husband, looking for more re-assurance. This time the hermit had scarcely seen them approach before he drew the hides across his face and muttered into his chest:

Alas, alas, the matter of my brother worries me, it
worries me. The figs are nearly ripe, ripe and ready to
fall. They will scatter across the ground. They are nearly
ripe, they are nearly ripe and they are ready to fall.

Again and again, until she drowned him out with her weeping.

The rush always began an hour or two before noon. After the long vigil he would stand on the steps of the church and say to everyone, Come. Come with me. Let us celebrate. Let us celebrate the day after which our church was named, the day when the child Mary came to the Sanctuary and did not turn back. And everyone took him up on it. Priest after priest; monks, nuns, deacons came stepping through their gate. Everywhere she looked were white turbans, yellow skullcaps, prayer sticks, horsehair fly-whisks. Woodsmoke and cooking smells began to mingle with incense and stale night-breath, dust and sweat.

In Addis the motor car might have been in the ascendant but in Gondar the grander people, the landowners, aristocracy, the wealthiest merchants, still arrived on brightly caparisoned mules, manservants running alongside, clearing the path. They were invited individually; it took three messengers a few days to get through the whole list. Her husband bowed and thanked them for coming and ushered them into the das. His face was composed but she knew there was pride and pleasure behind it; that his stance, outwardly deferential, rested on a foundation of defiance: See? Look who comes to my home. See how emphatically I have arrived.

In the das, which was now carpeted with rich rugs and meadow flowers and fresh grass, the chairs were arranged in order of hierarchy: antique wooden seats at the top for the nobility; somewhat lesser chairs for the head priests; carved stools for the common priests; benches for the deacons, the students, the male laity, and a handful of elderly nuns. On this first day, no other women would sit with the guests to eat. Yetemegnu remained in the kitchen, but she knew her husband would spend the afternoon circulating. And that considering the number of people, the rapidly depleting gans of beer, the steadily shrinking store of golden demijohns, it would be remarkably quiet as everyone attended to eating.

Another great wave rolled through their compound in the mid-afternoon, as the first shift departed and the second took its place. Not everyone left: the church students especially, young and spirited and far from home – they lingered, and drank, and as they became merrier and more crumpled began to test lines of qiné, flinging out rhymes and ditties, about the food, the beer, the mead, the holiday.

But eventually the shadows lengthened and the crowd thinned. She and her helpers began to turn their attention to the next day, and the day after, and the day after that, when more students, priests, relatives and acquaintances from throughout Gondar and the countryside, however poor and lowly, would come and bow and eat their fill. Sometimes it could take six, seven, even eight days before the flow of guests slowed; more before people stopped dropping by for a tipple and a chat – did you see who came the first day? I didn't think he could show his face. Wasn't that fish good? You really surpassed yourself! But at the end of the first day she loosened her girdle and sat down with a sigh, for the first time, it seemed to her, in weeks.

The man came through the gate bowing. Madam, how are you? How are the children? All well? Are you well? How are you? Well?

I am well, she answered, thanks be to God, we are all well, may His works be praised, and you? Are you well? Your wife well? But she felt his agitation, and she waited for the niceties to be over, until he could decently say why he had come, in the middle of the working day and in such a rush.

I have come from the prison. Yes, you are a guard there, I know, how is the work treating you? Well, madam, well. I have asked someone to stand in my place. I had to come because I am Gojjamé too and what they are doing is not right. It is not right. They have taken your husband, madam. They took him in the street. It isn't just him, they took all four from the Gojjamé association. Madam, if you bring food I will take it in, I will make sure he gets it. Madam, ayzot, don't be scared, please don't be scared, I'm sure he'll be out soon.

She saw as if from a great distance that she was wringing her hands and pacing and standing up and sitting down and wringing her hands again. There was screaming in her ears, too, and she realised the screaming was coming from her.

BOOK IV
1953–c.1958

TIRR
THE FIFTH MONTH

Height of summer. Some sluggish rivers considered poisonous for cattle. Harvesting of irrigated garden crops. Selling of produce to buy salt, chillis, coffee. Tax gatherers collect taxes for the government, for landlords, and for church tithes.

Nearly every day now she pulled her shemma over her hair, smoothed the fringed border carefully across her chest, indicated to the servant to follow her with the basket, and stepped out of the main gate. As she passed the two rows of rooms her husband had built and entered the street into the market she dropped her shoulders and lifted her head. Let them whisper. Let them stare.

If it was a Saturday, progress was slow. They would thread their way through tanners and drovers, blacksmiths and farmers' wives and all their customers: lean men peering into the mouths of mules for sale, women sniffing at spices, butter, Sudanese perfume, or making calculated flounces away from grain they needed, holding out for lower prices. On weekdays they could move faster, stepping across rocks worn flat by generations of market-going feet, scattered with kernels of barley, corn husks, chewed sugar cane. Dark crows pecking at the roots of the syca- more fig under which Aleqa Tsega had spent so many hours passing judgement.

Usually when she approached the tree she turned right, directly up the slope to the prison. Other times, needing succour, she continued along the brow of the hill to Ba'ata, bowing to kiss the dear walls before turning back. Oh Lady of Sorrows, help us. You who as a child were fed by angels in the temple, whose girl- hood and whose motherhood we honour in this place, help us, please.

At the prison door she tried to ignore the eyes watching from the sentry-house above her, waiting until the Gojjamé guard could come to her and she could hand over the injera and the chicken stew she now made almost every day. It was only a month or so since he had given her his news, but it seemed like years.

Every few days she was allowed in herself, to sit with the other wives, to chat as normally as possible about children and cooking and neighbours, to sift her husband's expressions for signs of ill-treatment and indications of mood. At first prison appeared to suit him: his face filled out, and his dark skin smoothed.

And they talked. Or he talked, always across the same ground, obsessive, outraged, unsurprised. They hate me. This is what they've always wanted. Yes, she said. They hate anyone from Gojjam anyway, but then I was promoted, and promoted again, and they had to answer to me. They're jealous. Yes, of course. I suggested the priesthood should be salaried; that no one should profit from church lands who wasn't trained, who didn't serve – that was a good idea. Well, they hate me for that too.

They say our little group of Gojjamés is a conclave of traitors, that I spent my days weaving plots against the emperor. She looked straight at him. No. They swore in court that when we help each other and encourage trade between Gojjam and Eritrea it's a pretence – that actually we agitate for federation and constitutional monarchy. No. When the judge sent them away for having no proof they went to the governor. It's Asratè Kassa, you know. He's back. The emperor has reappointed him. They asked him to take away some of my responsibility; he told me to share my duties with another priest, but I said my office was in the gift of the emperor, and I would not comply. That too will have offended him. They know what they're doing. But they are all lying. Lying before God.

She could say nothing. Her hand moved on the swell at her waist and she could not tell who was failing to comfort whom.

My lord!

She darted forward along with everyone else who had been waiting under the trees. Her eldest son leapt forward with her.

My lord!

She might as well have been a pipit dipping and twittering across the valley that dropped away from the church walls.

My lord! May the reason for my husband's imprisonment be known?

Asratè Kassa kept walking. He was a man of about her own age. Strong-featured, straight-backed, dark-skinned. Taller than most around him. A full moustache and a receding hairline. Utterly assured of his station.

In the name of the Saviour, may I speak?

Nothing. He was nearing the gate, and now she was wailing. Please, in the name of Hailè Selassie, upon his life and his death, may we know what my husband is accused of, that he may receive justice?

At last he turned and looked at her. Through her, it almost felt, the heavy eyes seemed to register so little of her presence. And then he turned away.

So her visits to the prison continued. Food, carafes of mead and beer, so he would not have to use the gourds that the guards provided for drinking, for eating or, in the reeking night, for other bodily functions. His litanies of accusation, counter-accusation, protest, self-pity, continued, held him in a grip he could not seem to loose.

The children are well? Yes. They're studying? Yes. From across the market came the emphatic rise and long fall of the muezzin's call to prayer. Allahu akbaaaaaar! – they cannot go to church school. They must not become priests or work with priests. Look what priests have done to me.

The initial hectic flush of health faded. One day, watching him cross the floor, and then cross the floor again, she noticed he was shaking. Sit down, master, sit down. He shook his head. She let him be. But it happened often after that. He ate less. He stood, and shook, all at once looking all of his sixty years, but refused to sit. Please sit? No, it hurts too much. Tears knotted in his eyes.

And then a day when he *was* sitting, on a pallet, pale. Yetemegnu, I can't stand. Here, let me help. No, no. The tears broke through. Let me help you up – no, I can't stand. He came to visit, and now I can't stand. Who came? That witness who testified against me. *That churchman.* And now my body is heavy to me, and I can't stand.

Oh my heart, she cried, ayzoh. What shall I do for you? What can I do? Then an idea so outlandish, so sure to be rejected, she spoke it aloud. Shall I go to Addis? Shall I try the higher courts?

He did not immediately dismiss it. She watched as it gained flesh in his mind. Perhaps – perhaps you should. Yes. Go to Addis Ababa. Go to the highest court. Go to the emperor. Tell him what has been done to me. Free me.

When she got home she could not be still for worry and fear. She paced through the rooms, picking things up, putting things down, kissing children, putting them down. How could she go? What about all these children? How could she go so far away, to such a big city? But she had to go, she had to free him. For him, for all of them.

'I have a relative in Addis, near their market, he will help you.' The neighbours had heard what she was about, and unable to resist the excitement, found reasons to drop by. 'You must take gifts, money. Everyone will want bribes.' Eventually her youngest daughter's godmother spoke. 'But madam, you must feed him yourself. Otherwise how will you know they are not poisoning him?'

But he said to go. I must free him. 'Don't go. Stay here. Look after him.'

She stayed away from the prison for three days, thinking about what to do. Then she pulled her shemma over her head as usual and stepped out of the gate. A man bowed to her from across the roadway. She had begun to respond in kind when she saw who it was and stopped, raw with anger. The man, who had testified against her husband in court, gestured at her. Stay, he seemed to her to be saying. You should stay.

She described the encounter to her husband. So they are going to humiliate me like that, are they? By kidnapping you, violating you, ruining you? How dare they! I have no doubts now. Go to Addis Ababa, immediately.

How can I leave you? I have to feed you! I have to make sure they don't poison you! I have to keep you safe!

Go! Now! Then, softening somewhat, Please go.

And again she said, Yes. Ayzoh. Yes of course.

She was still repeating it when she got home and began to pack. What would she need? Dresses, good dresses, ayzoh, I'll help you, ayzoh, I'll go, I'll go, headscarves, good headscarves, I'll go to the emperor and get you freed, her best shemmas, ayzoh, don't worry, I am here for you, still repeating it when she felt something snap within her, snap and then release, shocking her into stillness. And in the stillness she felt it begin, a soft warmth first, a gentleness belying implacability, and then the pain, arriving as if from a distance, coming closer and closer, then in waves, successive regiments assailing a fastness. Trickles became rills that coursed down her legs and pooled at her feet. The pain gathered itself, focused. She looked down. There it was, as deep a red as everything that surrounded it, all stomach and all head, limbs tiny as afterthoughts, stopped-up mouth and fused-blind eyes. A boy.

Everything about her was closing in, darkening. But the blood flowed on and on, through the receiving room and out of the door.

Later she was told that when she lost consciousness the women had panicked. They had tried to lift her, hauling her from room to room – 'Seven rooms!' they cried. 'It must be seven rooms!' – leaving blood everywhere, streaked across furniture, soaking through their white dresses, trailing behind them. They took up their pestles and beat them against any metal they could find, roof tin, barrels, pots. Someone rushed to the grain store, grabbed handfuls of linseed and threw them onto a pan on the fire, where the grains popped and hissed in demented answer to the arrhythmic crash of metal.

'What's all this?' It was the Muslim man to whom she and her husband had rented one of their new rooms. 'What's going on?'

'This woman is terribly ill, look, she's dying, we must scare the devil away –'

'Why didn't you call me? Now, go, bring me a new length of cloth, a piece that hasn't been touched by water –' He produced a long, narrow piece of vellum and leaning over it began to write. They stood back. Under his concentrating hands suns appeared, and crosses and moons; horns and burning eyes, hasty letters, letters and more letters. He wrote and wrote until the vellum was filled, then, rolling it into the fresh cloth, tied it securely to her left thigh. Immediately, the women told her later, the bleeding slowed. But it was a night and a day and a night before she woke, cursing the fact that she had not simply been allowed to die.

When she was able to stand, they took her to the prison to see her husband, who thought he had lost her. He wept to see her, and she wept, and they wept again when she rose to leave. They held each other for a long moment before she disentangled herself. Ayzoh. I am going now to free you, and I will be back soon.

In that country and in that time, when every thoroughfare still had its litigants, bringing earnest complaint before a kerbstone judge (often appointed on the spot for the purpose); where the ability to argue brilliantly for oneself in court was more respected than business acumen or craftsmanship or musicality; where justice was held so dear that most would countenance a lifetime of court appearances rather than feel it left undone; there everyone, however humble, had the right to bring his or her case before higher and higher courts, and eventually, if necessary, before the emperor, the final arbiter between them and God.

And so each morning she set off for the High Court, accompanied by Meto-aleqa Tirfé, a kindly lieutenant who it was said might have been a general had he not displeased his superiors. She was staying with another army man and relative of her husband's, an old Gojjamé commander. Basha Deneqè and his wife could not have been kinder, providing advice, a bed for Edemariam when he visited from boarding school, meals at all times of day, but she could not eat for nerves. A cup or two of coffee only, before wrapping herself in layers of shawls and going out into the streets.

Nearly thirty years of diurnal rounds consisting (except during the war) of her house, its grounds, the market, Ba'ata, and latterly the prison had fixed her horizons close. The skies felt different in Addis, the mountains higher and farther away; she felt somehow naked, and yet at the same time jostled and distressed, even when no one was walking too close. There were so many cars, honking, rattling over ruts, nudging through so many knots of people and mules, sheep, donkeys, that at any moment she expected one to creep up behind her and drag her under its wheels.

Within the court compound things were not much easier. So many people, all aiming for the same thing: a chance to put their story to the judge. They milled, and waited, and milled and waited, until waiting became a way of life. Every so often a name was called, an answer shouted, and the lucky person hurried up the stairs. By noon it would be obvious that her turn would not come that day, and so she would walk slowly home.

The sun was at its zenith. It blazed down on the rubbled asphalt of the roads, on the top of her head, on the umbrella she raised for shade. When she first arrived, still bleeding, breasts hard with milk, the rainy season had not yet ended, and so she had often glanced up to see an advancing wall of silver-grey raindrops falling straight and heavy as spears, churning the dust before them into crowns of mud. When the clouds had passed, as swiftly as they had come, the ditches would run with white-pointed brown torrents and the flotsam of the city: sandals made of car tyres, sodden paper, vegetables, rags, bones, excreta. The roads would steam in the sunshine, the birds singing out, her nostrils filling with the smell of wetted earth, of bruised green and mountain air.

Now, in the afternoons, she began to pursue other avenues of redress. She recalled her father's interminable counting of houses, that she was distantly related to the empress, and thus to the empress's eldest daughter, Princess Tenagnewerq. She knew that of all the emperor's children it was probably this daughter, not the crown prince, who had the most influence on their father. Moreover, the princess's husband had only just been transferred to Eritrea after six years as governor of Gondar. He had met her husband, had dealt with him officially.

Of course, she also knew other things. She knew the couple, like much of the royal family, had business interests everywhere. She had heard whispers, she had no idea if they were true, about his roving eye. But mostly she remembered that in the free-for-all that followed the Italian retreat, the princess and her husband had acquired large swathes of land in Gondar city. That some of that land belonged to her own father; that her husband had asked for it back and had instead been offered forty hectares nearly five hundred kilometres south, in the Rift Valley. She had heard it was decent land, arable, but what in the name of the Trinity were they supposed to do with it?

But she had to set that aside, for the moment. Or perhaps here was a way in which she could be repaid? So she began to walk through the lunchtime streets, and to take up a position outside the high stone walls of the princess's residence. She was not the only person to have this idea, but then she would never have expected otherwise: every great man or woman in the country trailed a penumbra of petitioners from all walks of life, often felt harried by them, but would be outraged, take it as a slackening of power and status, if ever they melted away. And so she took up a place among everyone else who stood, or sat, or paced, waiting for a pause, a glimmer of permission to speak, for hours and days and weeks.

On other afternoons she went to see her husband's friend the bishop, and this she much preferred. Abunè Theophilos was a busy man, increasingly powerful, concerned, she would much later discover, with reorganising how the church was run, shoring up its ability to help the stricken, a supporter of modern education and a scholar in his own right. But he found time for her, leading her to a seat under a grapevine in the garden, asking after her health, her children, her husband.

No change, she would say. No change, may the Lord bless us and keep us. But then, one day, I have had a phone call, they have taken him away! They have taken him away! Where? Please try to be calm, my child. Where have they taken him? They took him in chains! Soldiers took him, in a big car! – to Wegera, in the Simien mountains, where he was now under house arrest. She could not understand. Was it because the governor had heard she was here, trying to go above his head? Did he want to make sure he didn't lose his prisoner? Had there been pressure from Addis? She could not know. All she knew was that Wegera, set amongst some of the highest mountains in the country, was remote and at this time of year, when dry-season sunshine gave way to clear-skied nights, laceratingly cold. She also knew that her husband was increasingly unwell, that he was bleeding, awful private bleeding –

'Ayzosh. Ayzosh. There will be an answer. He will not be there long.' She looked up at him, wanting to believe. Abunè Theophilos was tall, handsome, with a long nose and slightly hooded eyes, high cheekbones swooping down toward a short squared-off beard. Two women of his household sat a short distance away, looking askance. She saw the question in their faces – what was he doing, talking for so long to this Gondaré girl, so thin and weak from childbearing and breastfeeding and weeping and not eating that her hips jutted out against her dress like guns?

The next time she came she was even more distressed. Her husband was sicker. He had been transferred back to Gondar, to the hospital. He had been placed first in an open ward, but the hospital director said he could not allow it, not for a man like Aleqa Tsega, and so he was moved to a private room. He would not eat, though that was more from anger than from illness: he was consumed by anger, riven with it. Everyone who visited brought food, as tempting as they could make it – buttery gruel, porridge, rich stews – each dish tasted by someone else first, to set his mind at rest. His two youngest sons were often at his bedside. But still he would not eat.

She had been going to court every day, she told Theophilos, desperate to move things on, and she had had some success – she had been called before the judge not once, but twice. The first time she was trembling so much with nerves the judge had had to rush from behind his desk to catch her before she hit the floor, lifting up her chattering bones and setting them down on a chair. Both times she had eventually been able to speak for herself, and two separate letters had been sent to the governor in Gondar, ordering her husband be brought south and tried in the capital. But so far there had been no response.

Once, sitting under the grapevine, emboldened by his kindness, she spoke of herself, as if she were at home and at confession.

Father, I too am ill –

In what way, child?

I am visited. The zar comes – I drink coffee in the morning, when I ought to be fasting. The zar requires it.

Ayzosh. My mother used to suffer the same.

The next day a huge sack of coffee arrived at Basha Deneqè's gate. Speaking of this over half a century later she remembered what became of Theophilos, his awful death, and cried out in sudden anguish, hands raised to the heavens: why, why, why?

When she took her leave, bowing to kiss his cross, bowing to the ladies, and went out into the street, her steps were lighter, her head a little higher. The sky was hazed with dusk and smoke from evening fires. Children called, their voices tinny with clarity and distance, dogs wheezed and barked. And above them all a lone kite sailed, at ease on invisible updraughts, impossible wingspan utterly still.

Within a week word came from the bishop's house. The emperor is at his court tomorrow. He will see you.

She hardly knew how they got there, hardly knew how they passed through the blur of uniforms and capes, gardens and crested archways. Up steps. Along cool halls.

A large room and at its far end a small figure – how small! – standing. Head almost too big for narrow shoulders. A long flared nose. A wide brow scrawled with lines. Sunken cheeks. Did he not sleep properly, or eat?

Instantly she directed her eyes downwards, at the carpet, her head and all her body bowing. When she was told to speak she did so without looking up.

His voice, when at long last it came, had a displeasing, shredded quality. It was as noncommittal as his bearing.

'Let him come here. We will hear the case.'

*

A smoky room and a terrible blind dance. Dancing and dancing and wild harsh voices, full of malediction and prophecy. Eventually the tongues ceased and the women looked up. Her hair had been dragged out of its scarf and stood ragged and torn from the back of her head. Dread tightened their chests and caught their breath.

Sleep, deep and dreamless, for a day and a long, long night.

Another room. She had been ill for two weeks, unable to eat, her insides water. They fed her sunflower seeds, thinking these at least she could keep down, but they only made her worse. When, finally, she began to improve, a man came to her, a relative, and she sat, weak, to receive him. How are you? Well, thank the Lord. And the children? They are well, may His honour increase. A long pause. Your husband is ill, madam. Yes. He is ill and he has been released. Released! Oh, thanks be to God! And you must go, now, to care for him. Yes, yes of course. Thank you, for bringing such good news. You will excuse me – And she called servants to her, and began to pack.

A room, and Edemariam, dreaming. He dreamt of his father, who turned, and walked away.

They woke him an hour later.

Rivers shrunken, sloughed snakeskin laid down in hollows across the brown land. Flat-topped acacia cresting the near hills, and euphorbia, and in the distance, climbing into hot blue sky, the

mountains she had always known. The horse's hooves clopped
steady over the dry wheel-ruts in the airport road, and the garri
jounced in response. After four long months she was going to see
her children again. Her husband, and her children. The children
must have grown.

There was the gate. Quick, open it.

She lifted her skirt to her calves and stepped through, into her
own compound.

A das. Why a das? *Why a das?*

'NOW WHEN THEY SEE ME COMING TO THE TOMB AND
PRAYING, THEY REVILE ME, AND MAGNIFY THEMSELVES
AGAINST ME.'

– LEGENDS

Oh my people why sit so silent?
Let us weep together.
There is not one among us
Who has not lost a brother, or a sister.
Oh let us weep together.

The mourner's voice cracked into tears, successfully taking everyone else in the das with her. For a long moment they sobbed and
murmured before lapsing again into quiet. The das was almost
full. For days she had sat under the eucalyptus joists and boughs
as people arrived, to sit with her for a while, to weep with her, and
then to leave. Every so often a verse would arise, from a profes-
sional mourner, from a gifted relative, from anyone who felt so
moved.

Who says Bitwoded Tsega is dead?
We saw him with Abraham in heaven.

And the weeping would begin again. When she first arrived and understood her husband was gone she had been insensible for three days. Now she was largely conscious, but inert. She picked at pieces of bread, turned crumbs round and round in her mouth, struggled to swallow, failed. She felt light-headed, and the days were unending, or suddenly over, twisting and passing in ways she could not understand and did not have the energy to follow. The nights, spent on the floor with her arms around her children, were full of tears and bitter prayer. Her eldest son had had to stay at school in Addis. The two younger boys, who had watched their father decline in hospital, now followed her with serious eyes. The girls clung, the baby was ill – her belly was puffed up, tight, while the rest of her was bones, poking through a bag of skin. Yetemegnu's voice rose, hoarse.

He was a man, but after I gave birth
He cooked meat for me,
He did not let me sleep.
Caring for me like a woman, day and night.

Alemitu was exhausted and overwhelmed from running the house alone. And she had her own griefs. When her mother had left for Addis Alemitu was still producing milk, so she had fed both her own child and her youngest sister. Then they had all caught whooping cough. Eventually seven corpses were carried out of the big house – relatives, tenants, and Alemitu's small son. Her youngest sister would feel guilty about surviving well into adulthood.

The eaves of my master's house are leaking,
The rain is pouring through.
It has not been standing for long
But already it is falling into ruins.

The mourners attempted comfort. He was a great man. He was well loved. She had her own opinions on the latter, but for the moment she held her peace and listened while they told her of the funeral she had missed. Rites were read for him in the forty-four churches of Gondar. So many people came – Gojjamé soldiers from the barracks at Azezo, Muslims from Addis Alem, civil servants from offices closed for the occasion, priests from throughout Begemdir and Semien – that the funeral had to be held in shifts. Ba'ata was full. Ba'ata's grounds were full. The streets between Ba'ata and the house were full; there was nowhere to put one's feet. People spilled out of their compound and down the slope into the Saturday market. Even the governor came. When he arrived, they told her, her half-sister had looked straight at him and spat out an old verse.

They hacked at him, sawed at him
As if he was an elephant.
Look! There! He's felled for you now
If you want to consume him.

It was months before she was told that at his sickest an order had come to the hospital, to let him go home. But he had flatly refused to do so until it was explained, officially and in public, why he been imprisoned in the first place; until it was proclaimed that he was innocent. That in his last days her half-brother Gebrè-Selassie, now a tax collector in Dembiya, had said to him, 'Your wife should be with you now. Let her return from Addis and care for you.' That her husband had snapped back, 'Don't be in such a rush. You'll be able to marry her off to someone else soon enough.' And Gebrè-Selassie, disappointed and angry, had in silence turned away.

I will never, ever divorce my husband.
My husband, who when I was a child braided my hair,
Trimming the rough edges, teaching me manners.
My husband, who raised me.

She watched everyone, all the time, feared all but the closest to her, feared poison especially. The children were no longer allowed to eat anything but food supervised by herself or Alemitu, in their own home.

Where is he? she said to Gebrè-Selassie one day, all of a sudden remembering the wild teenager who had leapt to the defence of his mother all those years ago, nearly killing all of them. He had ranged through the hills as a bandit for a while, and now spent his days hunting game. Could he help me? If I pointed out each false witness? For days she hovered on the edge of sending a messenger, but never quite shored up the courage.

Please bring me a flat basket
With which I may winnow the earth.
All that learning, all that intellect,
Where was it spilled?

Some mornings she woke at cockcrow – some mornings she hadn't slept – and slipped out of the house. Habit directed her feet north-west, toward Ba'ata, but she refused habit and instead turned north-east, toward the castles and Gimja-bet Mariam. Under the cedars she bowed and tried to pray, but her mind could not let go of the other church – he suffered for you, he carried rocks, he laid your foundations, you were his life's work, and my life's work, everything we did was for you, why did you kill him? Why? Of course she knew it was not Ba'ata that had killed him; betrayal had struck so deep, however, that it did not matter. Could you not have let him finish? I will never kiss your walls again, never, do you hear? I will not bow to you, I will not bring my children to you, you are dead to me.

> Honourable Aleqa Tsega, favourite of governments,
> Beloved of God and friend of the people,
> He is still to be found
> At the entrance to the sanctuary.

Because he had made her swear she would never do something so financially ruinous she did not hold the traditional memorial banquet, but at forty days his relatives killed an ox and fed the priests. She was not required to cook and so she sat and watched and thought, How can you know your real friends? How can you ever know?

> Oh all you people, let me tell you,
> Before a judge, before the saviour,
> What will become of me.
> Sorrow does not kill.
> If sorrow killed, how willingly I would die!
> And yet still I live, burning with fire.

Gradually she began to disentangle their various sources of income. The rooms her husband had constructed along the outer wall were rented, and so there was money from that. There was wheat and barley from Gonderoch Mariam. The Jews who farmed it had recently objected to giving up a third of their produce each

year, so the tribute had been reduced, but was still ver
land in Dembiya belonged to the church, of cour
reverted to the church, but they still had Bisnit. And Bisnit was
beautiful land, nearly a hundred acres rustling with green in all its
shapes and shades, cabbages and kale close to the ground, false
banana leaves like ripped flags, darker bushes leaning over a
spring that spilled into a creek that at this time of year took barely
a stride to cross. Each harvest season they hired two oxen to
plough it, but it was too big for them, and so her husband had
invited another man, a merchant, to farm it with them and share
the proceeds: teff, linseed, peas, beans. More recently they had
decided to plant trees on it, a thousand valuable eucalyptus. Her
husband had suggested the merchant's daughter marry their
eldest son.

But when, thinking that these trees might pay for her children's
schooling, she began to make enquiries, she discovered that as
soon as her husband had died a group of monks from a remote
northern monastery had claimed his share. And that they had
been negotiating with the merchant, who had promptly passed on
the half she thought was hers. She was, at one level, unsurprised
– was she not a defenceless woman, sheltered and easy to take
advantage of? But she was also entirely outraged. How could they
do this to a widow with seven children?

> Oh remember Master Aleqa Tsega,
> Recollect his good character.
> Where are you going,
> Now that your children are growing up?

There is a man from the Sudanese borders, they told her, who
specialises in casting the evil eye. He will lay waste to your
enemies.

And in her grief and anger she sent for him and led him into the little alcove at the back of the house. She poured him a horn of beer, and asked him to be seated. Eventually she sat too, and among the grain and onions, under the strings of drying beef, spoke nervously of this and that. How was your journey? Do you have family? How are your children?

They spoke of her own children – her oldest son in Addis, the boys who were still at home, her daughters. And then, in a rush, of her husband, how he died in agony and of those she thought – she *insisted* – were responsible. She could not bring herself to say exactly what she had in mind, but she looked at the man intently and said, Do – what you do. I will give you eight of my rental rooms if you … do as you would do.

He sipped his beer and looked at her. And she looked back, shaky, defiant. He said nothing. The quiet stretched, solidified. A rooster crowed in the garden, and children scuffled. But in the little room there was only silence.

At last, when she thought she could bear it no longer, he spoke. Madam.

Yes?

Aleqa Tsega is in heaven. He has been sanctified. We all know how Christian he was. And you – by association with him you are the same. You are Christian too, and your soul is still pure. I cannot do this for you.

What? she started. What? I asked you here – but even as she began her words emptied of conviction, and the tears arrived.

He saved my soul, she told her children when they were grown, part-laughing, because her behaviour had been so entirely uncharacteristic, but part-serious too. I thanked him, because he saved my soul.

It doesn't seem so to him,
But the harm done to one hurts the other.
The pain of the retainer
Diminishes the king.

Some time afterwards there was another visitor. A man she knew this time, a careful, short-sighted civil servant who had risen to be secretary-general of Begemdir and Semien.

After the many conversational necessities – and this very proper man would always observe every single one – there was a pause, and then he said, 'You know the governor –'

The mortification of her encounter with Asratè Kassa was as intense as if it had happened yesterday. Yes, I know the governor.

'I was there, madam, when the letters arrived from the courts and from the emperor's offices, ordering that Aleqa Tsega be brought to Addis Ababa to be tried. I watched the governor open them, and read them.'

Yes?

'He ripped them up, and threw them to the ground.'

This time she left for the capital laden: grain and coffee and spices, scrubbed white clothes. And two of the three littlest girls: the youngest, now nearly a year old, and Tiruworq, not very much older, to help look after her.

A small place had been prepared for them at the home of her husband's first teacher, Memhir Hiruy. He lived, now, in the palace grounds, and had given his home over to another priest, a bent-over little man who had served at Ba'ata, and had often been a guest in her mother's house. When this aleqa was not at the palace himself, manoeuvring, he ran a church school in the compound, and so she set up home to the fluting sound of small boys reciting psalms.

Night-time was a different matter. The compound was near Trinity Cathedral, with its pale bombastic arches and dark onion domes, and to the parliament building, where members debated constitutions they knew would almost always be overruled, but it was also close to a vast slum, known to all as Erri bè kentu, or A Cry for Help in Vain. When the sun had set and the main roads held only scavenging dogs and the occasional loping hyena, they could hear it: erratic music, laughter, shouts, crashes, screaming, and then bated, waiting silences. She hugged the girls close to her and prayed aloud, a separate prayer for each individual child, for her dead husband, for herself. For the swiftest possible justice.

Everyone had advised her not to come, especially when she told them exactly what she wanted, which was not just the land, but the clearing of her husband's name, for his sake and hers but above all for the sake of her children; she wanted blood, and as the blood she wanted was Asratè Kassa's, she intended to ask the emperor for it. You can't do that! Quick glances, around the room, out the window. Everyone knew the emperor had spies, and not just in important places. If they find out they'll kill you! She had listened, polite, then proceeded with her preparations.

Because her nights were restless and spent in prayer and weeping, weeping; because in those sleepless hours she rehearsed the endless steps she would have to take, the favours she would have to beg, the days and months of waiting and contention, and she was afraid.

Already men were circling, eyes not just on her land but on her person as a candidate for marriage. Considerable men, qengaz-matches, minor royalty, one or two of them handsome (and she had always appreciated a handsome man), others promising to set her aflame with gold.

But how could she give in? How could she bring another man into her home, to watch the hands of her children as they ate? To insist on offspring of his own? – for of course a good part of her attraction was her abundantly proven fertility, and though she was now in her mid-thirties children were still possible. The children she had already would, in another man's house, inevitably come second. Why would she do that?

All her life she had listened, rapt, to the lives of the female saints, of Zenna Mariam, Fiqirtè Christos, Weletè Petros – fierce women, many of them, beautiful and well-born, pious mothers who left their children and the world to become nuns. And especially Christos Semra, mother to eleven, who after she became a nun at the monastery of Debrè Libanos submerged herself entire in the waters of Lake Tana and prayed without sustenance for eight years before rising onto the island of Gwangut, where she founded a convent and was vouchsafed many miracles and visions.

She remembered the vows made in the church when she was too young to understand: I will marry this man, and this man only, and if he should die before me, I will renounce the world.

For years she had looked at nuns with something approaching envy, thinking how their yellow skullcaps signified virtue and removal from the world, but also a kind of licence. And after the years of domestic labour, of privacies denied and advances inescapable, of the bearing of children, the suckling and the carrying and the worry, that seemed like light and freedom to her. In the priests' wives who had remarried she had seen how her life could go: their houses were poor, their tables bare, and those who had become nuns – how beautiful their houses! How calm!

She had her inheritances, the lands and the house and the self-sufficiencies they represented. Marry once, and leave it be.

For days she sat, thinking over all these things, and then she called her confessor to her and together they boarded a bus north, to Debrè Libanos.

By the time they reached the perimeters of Addis they were already climbing through forest. The eucalyptus trunks stood smooth and pale against the dark of the junipers and cast shadows like bars across the narrow road, making the early-morning sun flicker and refract across their vision. The earth was red underneath the trees, rust-red and bare rock, and where years of rainy-season storms had washed the soil away there were small cliffs and crevasses bridged by ladders and lattices or the single gnarled limbs of exposed roots. Deep in the valley, falling away to their left, corrugated-iron roofs winked at the sky. Blue woodsmoke blurred the horizon.

And then they were no longer climbing, but bustling through open fields. There were still high peaks, but they were far away; here the land stretched itself out under a bowl of sky. Shepherd boys moved tiny across it, and wobble-humped zebu; egrets stood thoughtful, implausibly white, in the reeds at river bends. Goats chewed, plotting delinquency. Eucalyptus saplings glowed in the copses, while all around them fields unfurled, golden seas punctuated by domes of gold. The sun glanced off each blade of hay and set the stacks alight.

This, some of the richest land in the empire, was Kassa country, the base from which the old lord went out to fight the Italians, and where Asratè Kassa was born. These farmers they passed, who spent backbreaking weeks guiding heavy oxen and recalcitrant ploughs across the earth, who sowed and harvested and threshed and winnowed, standing by piles of grain higher than themselves and throwing it aloft until they were surrounded by aureoles of dust and chaff? Their taxes went to Addis Ababa now, but for generations up to a third of their yields had gone to Asratè Kassa's forebears. She watched the farmers' homesteads go by, low clusters of houses inside acacia fences aggressive with thorns. The thatch shone too, deep and dark as a gelada's fur.

And then the land ended.

Before them, where there had been track, was only blue air. There was nowhere to go except right or left, and so the bus bore right along a gorge which dropped and dropped until the braided channels of the river at its base seemed delicate as threads pulled from the hem of a new dress. Kites drifted below them, and snatches of cloud. Lungs filled with space, and an involuntary awe. On their right another cliff rose, and so they were travelling, now, along a kind of shelf.

They began to pass other pilgrims. Women carrying umbrellas against the sun. Monks in black soutanes and turmeric-yellow gabis. Nuns, yellow-robed, yellow-capped, chests armoured in concentric rings of heavy amber. A market under a fig tree, neat little piles of incense, limes, raisins, on small squares of cloth laid on the ground. Lepers crawling up the incline on knee-stumps wrapped in dirty rags.

By the time they arrived at the church there were people everywhere, under the trees, among the graves, people praying, chatting, reading, sleeping. And women, many women: becoming a nun at this most holy of monasteries demanded a minimum of three, often seven years of daily service. After she had kissed the church walls she and her confessor left the grounds and turned onto a rocky path that hugged the base of the cliff. Soon it began to rise into a cool green tunnel. Vast old trees leaned overhead, the higher branches laced with vines, the lower, thicker branches silken. White butterflies danced through the sunlight that threaded through the canopy. The air smelled of mint and acacia, and hummed with bees.

She heard the first holy spring before she saw it, dribbling into the back of a crevasse tucked under the base of the cliff. It was far quieter here, the only sounds the water, dripping onto basalt crazed into diamonds, pentagons, elongated octagons, and the precise tick of dry leaves landing on rock. After she had been blessed they moved on, picking their way across the tumbled boulders of a pool, empty now, but in the rainy season churned by a high waterfall, and impassable. The cliff was riddled with crevices and caves. Hermits lived in some. Others were piled with the bones of generations of pilgrims who had come to be buried here, and, more recently, the bones of monks the Italians had asserted were accomplices to their viceroy's would-be assassins. All of Debrè Libanos's 297 monks were shot dead and the church razed; two thousand people were slaughtered and a further thousand sent to concentration camps.

The second holy spring had been directed into a pipe that led into a small concrete enclosure and bifurcated, a tap for men and a tap for women. Across the path was a low tree-sheltered area, where Emperor Hailè Selassie, like Menelik II before him, came to take the waters. It was modest – even the most ostentatious of kings knew this was not a place in which to parade his wealth – but there was nothing modest about the view. After many wanderings, said the monks, the archangel Mikael had brought St Teklè-Haimanot here and said, look down. Birds darted, glided, hovered below him. Count them. But he could not count them. These are your spiritual children – they are innumerable. A cloud descended and filled the valley. And this is the Holy Spirit, said the archangel. It covers everything.

The saint was said to have stayed for the rest of his ninety-nine years, founding the monastery and taking up residence in a cave that reached deep under the mountain and echoed with water. For twenty-two years he stood praying on both feet; for the next seven he prayed on one, until the other leg withered and dropped off. Here, in the forecourt, next to the cave, was the holiest of holy water, coveted by believers for over six hundred years. Reverently she joined the other pilgrims in their milling nakedness, waiting her turn.

The piped water hit her back with the force of a blow. She moved further in, welcoming it. Another blow.

'Wiy! Wiy! Wiy! Ay Saytan! Ay Saytan! The devil! The devil has come!'

Fear scraped through her. What was it? A snake?

No, blood, tendrils of it, mingling with the water flowing across the rock. Had someone hurt themselves? That didn't explain the horror. Menstrual blood. It must be menstrual blood. Absolutely no one was allowed in these holy places if they were bleeding.

'Come, enaté,' said her confessor. 'Come, mother, come out of there.'

Why would I do that? I want to be blessed.

'Just come out, my dear.'

Reluctant and annoyed, she moved away from the others. And felt it, moving down her leg.

How could this be? She had waited for two weeks after her last flow, had washed carefully, purified herself as well as she knew how – she could not understand it, how could it be? How could it be? And the desecration of it –

'Come with me,' said the priest, moved by her misery. 'Come with me.'

When she was clothed he led her down, back into the valley, and introduced her to an old monk. 'He carries St Teklè-Haimanot's gold cross. He might be able to help you.'

Father, she begged. Father, bless me, please.

The monk held the cross toward her. She kissed it, again and again.

Father, I came here to be a nun, and I thought I was pure, but I bled. I thought I was clean.

'Ayi,' he said, sorrowful. 'Ayi. Never mind. It is the devil's fault, not yours. Take courage.'

But I am also sick, she added, I am sick with the zar and it will not leave me. How can I be rid of it? Will it prevent me from receiving God?

'There is nothing we can do about the zar. You must placate it. And no – it will not stop you from receiving God.'

He turned to her confessor and handed him a small pot. 'If she came all this way she should not stop. Keep her away from the spring, but bring the holy water to her in this.'

When, eight days later, the course was finished, they returned to the monk.

Father, I am pure now, and I look forward to becoming a nun.

'You receive His body and His blood, don't you?' he replied.

Yes. Yes, I do.

'Good. So why are you so concerned about a bit of cloth on your head? Come to mass this Sunday. There will be a vigil before-hand; join that too.'

And after that?

'Be a good Christian. Keep yourself separate and holy, and continue to take communion. And go home in peace.'

Awash with fury and dismay she bowed and kissed his cross, and did as she was told.

*

In the mid-1880s King Menelik, named after the progenitor of the Solomonic empire and determined to expand that empire as far as his diplomatic abilities and the might of his armies would permit, had recently remarried (his second union, and her fifth) when the royal household picked their way down the slopes of Entoto and pitched their tents on the plateau below. The land was lush with grasses and flowering trees, but most important to a king who suffered from rheumatism were the hot springs that bubbled out of the ground and seeped, sulphurous, into the red earth.

His wife Taitu stood at the door of their tent and looked out. She was formidable, a woman who beyond her already rare accomplishments – the ability to read and write Amharic, to play chess and the lyre, to compose poetry – was decisive and forthright, physically brave, and gifted in the oblique arts of acquiring and keeping power. 'May I build a house here?' she asked. Menelik had no objection; remembered, in fact, a prophecy his grandfather had made, that one day there would be a city in this spot. The next year, while Menelik was in the south, overthrowing the emirate at Harar (and acquiring, thereby, a remarkably useful economic base), Taitu made the prophecy reality, moving down the mountain. Remembering the cascades of lemon-yellow mimosa blossom that had surrounded their tent she named the new city Addis Ababa, or New Flower.

Three years later Yohannes IV fell in battle against the Mahdists at Metemma and Menelik became emperor of all Ethiopia. He insisted the coronation be held in the church his queen had established at the top of the mountain, but it was a last hurrah. A palace already existed on the plateau below, with neighbourhoods emerging around it according to the configuration of a camped army. It burned down three years later but Menelik II simply built another, on a brow of land overlooking the steaming springs: a three-storey living complex ringed with balconies and topped at one end by a dome; kitchens, stables, embroidery rooms, smithies, arsenals, a mint, anteroom after anteroom, and a vast reception hall each of whose three gables boasted a row of fifty ostrich eggs. Indoors, under electric lights and stained-glass windows, soldiers, peasants and priests could feast on raw beef and beer at a rate of nearly seven thousand guests per sitting.

Four years later a hundred thousand fighting men – and a roughly equal number of camp followers – marched north with Menelik II and his empress, who personally led five thousand infantry into battle. She also organised over ten thousand women, ensuring they carried drinking water to the troops, and directed, with her husband, the building of a defence perimeter. Their decisive victory over the Italians at Adwa sealed the importance of Addis Ababa as the first fixed capital of an independent Ethiopian state since the collapse of the Gondarine kingdom over a century earlier.

Menelik II enjoyed their achievements for only a decade: he suffered strokes and declined into paralysis and senility, while his wife, feared and eventually outmanoeuvred, was confined to his sickroom and then, when he died, sent back up the cold mountain to live out her days in a modest house in the grounds of Entoto Mariam church. Menelik's heir, Iyasu, spent little time in his grandfather's palace and in any case was deposed within three years by nobles and clerics appalled, they said, by his favouring of (or, attempts to bring equality to) the Muslims of his empire. He was replaced by Ethiopia's first empress, the far more biddable Zewditu.

Zewditu's modernising successor Hailè Selassie I built his own home as soon as he decently could: within four years of his coronation work began on a new palace, up the hill from the old one. It was finished only seven months later, in time for a visit from the crown prince of Sweden. The old ghibbi bustled still, lions paced their cages or sprawled across the entrance to the throne room, where visitors who could bring themselves to step past them could inspect gold pillars, Persian carpets, and Menelik II's throne, hung with silk. But in many ways the centre of power had moved, to a cool European-influenced two-storey building surrounded by gardens. Hailè Selassie I called it Genetè Le'ul, Paradise of the Prince.

At dawn each morning Hailè Selassie I rose from his small bed with its crowned canopy and embossed coverlet of silver-blue and went to his desk, to read, to write, to think about the day ahead, although sometimes there would have been so many petitions, so much work to get through, he would not have been to bed at all. By 7 a.m. he was at prayer, and then at exercises – any grandchild who burst in at this hour received a good scolding – before breakfast at 8.30 and the beginning of his public day. If he looked up he could see the tops of the mature trees that now shaded his gardens. And he could see the spiral staircase not far from his front door. It had no use as a staircase, was not attached to any building, it simply climbed into the sky. The Italians had built it, one stone step for every year of Fascist rule, beginning with the March on Rome in 1922: when Hailè Selassie returned he had claimed the structure for himself by placing on the highest step a stone lion. He looked down again, and continued to work.

Down at the bottom of the hill, roughly halfway between the old palace and the new, Yetemegnu woke too, bullied toward consciousness by the hard earth beneath her hips. When her husband died she had felt she had to follow all the traditions of deep mourning, shaving her head, eschewing the comfort of a bed. Much later she would wonder what it was all for. No amount of self-abnegation – and she had slept on the ground for a full three years – was ever going to bring back the dead. Carefully she rose, wrapped herself in warm shawls, and let herself out into the dawn.

At this hour the streets were mercifully clear. The occasional stray dog, perhaps, nosing through a rubbish heap. Nightwatchmen. Country women bent under wide loads of firewood. Were they widows too? How many children did they have to feed? The cathedral grounds were quiet, but not unpopulated: lone women approached its walls, kissed its thresholds in private supplication. Lone men read under the trees.

When she returned the sun had crested the mountains and the air was warm. Her daughters were awake, and the compound smelled of roasting coffee. She sipped three small cups, saying little, eating nothing, waiting to give the prayers of benediction, which she said quickly, intent and musical, hands held palms-up to heaven, before standing to leave the house again. 'Be careful!' called her Eritrean neighbour. Sometimes, when this neighbour felt worried about the children, left to fend for themselves, she brought them fresh bread, hugged them. 'You're always crying,' she said to their mother. 'I'm surprised you can see. Watch out for the cars!'

Up the hill, past the square with its curving bank and facing post office. Past the leafy grounds of the archbishop's residence. St Mary's, his church. The Supreme Court. Into a smaller square, where the victims of the Graziani massacre writhed in bronze bas-relief across a needle of white marble. Through the vast stone gates, after some negotiation with the guards, a now-practised listing of names and connections, or a surreptitious exchange of coins. Up the wide avenue, flanked by lawns and flowers and trees. She was part of a flow now, all tending toward the same point: ministers and sub-ministers and sub-sub-ministers who knew that basic self-preservation required them to bow to the emperor,

and for the emperor to see them bow, and for their colleagues to
see the emperor see them bow, each morning before work.
Petitioners like herself, who also came nearly every morning,
sometimes for years. She looked up, searching above the heads for
the lion on his airy steps, and made her way toward him. She
found a patch of shade, and then she settled down to wait.

Sometimes, when she knew the emperor was not at home, when
she had seen his sleek car glide away through crowds that bowed
as it passed like grass bending away from a big cat hunting, she
turned and made her way down the hill again, and presented
herself at another great gate.

She was almost grateful for the familiarity of this particular
wait. It wasn't long before she had managed to deputise a kindly-
looking soul to go inside and wait for her, in case her name was
called: the court anterooms were so crowded they made her fear
she might faint, or lash out. In fact she already had, once or twice,
done exactly that – hemmed in under the obligatory portrait of an
impassive emperor, pushed forward by impertinent, impatient
men, she had turned and slapped their hands away. Stop! Stop
now! What kind of woman do you think I am? Outside there were
nearly as many people, but at least there was air and light and sky.

Sitting in the shade, fanning herself occasionally, she tuned in
to the chat around her. Who was hopeful, who had had their case
knocked back. Whose relatives were proving difficult – they are
impossible, may Hailè Selassie die if I lie! – how exasperating it was
that no one could get any justice at all these days without money
changing hands. There was much shaking of heads and sucking of
teeth. 'May the almighty come to your aid. May He guide you, and
keep you.'

But she could not sit there all day either. The central thing was to get to the emperor. He might even be able to – he *had* to be able to, he was the emperor, after all – solve her land disputes as well. So she left, and continued down the hill. By this time it was afternoon, when everyone, it sometimes seemed, was out on the streets. And a capital's streets were so different from those of a provincial town like Gondar, however overweening the latter's self-opinion. So much disease, displayed for gain. So many young boys, younger than her son, gathering at street corners, offering shoe-shining, boys from the regions sent to better their families' fortunes, boys carrying so much hope on their bony shoulders; hope, if they had failed the grade eight examinations in the new schools or knew no one influential, that would probably just wither away. Modernity then aped tradition, sending them out, like their fathers at church school, to beg for meals. But it did not offer tradition's safety nets, or any concrete promise of graduation to a different stage. She thought of their mothers and the tears welled. There were knots of donkeys, as always, and milling flocks of sheep. She watched as farmers picked choice specimens up by their front legs, or grasped rams by their curling horns and dragged them unceremoniously toward prospective customers. Mules, too, but this was less common than it used to be. They were outnumbered by cars now, driven by earnest young men in Western suits, or even, sometimes, by a woman.

There was something else in the air too these days, a kind of inchoate hurry. She began to notice just how many buildings she passed were obscured by scaffolding, how many were half-painted, or half-built. There were labourers everywhere, leaning on pick-axes by piles of rock or cement, waiting – for what? Nothing probably, just the end of the day, or the unwelcome appearance of a shouting foreman. They would certainly scramble into action then. She recognised the tendency from all the years of her husband's church-building. Arches rose over the main avenues, all wonky eucalyptus struts and loud lettering.

At the princess's spiked gate she no longer had to queue with everyone else. The first time she had been ushered in by a smiling boy who seemed to expect her she had been surprised, but it transpired that it was the empress who this time had counted houses and had said to her daughter, here is a relation in trouble and alone in the world. Look out for her. So every few weeks – for she did not want to abuse the privilege given – she was ushered into the princess's quiet living room, bowing, attempting to erase herself even as she hoped to be noticed, speaking quietly when the princess asked after her, after her children, after the progress of her case. The empress's directive had been handed on to the household too, so that even if the princess was absent she was led through a side entrance into the house, where she perched tense on the edge of a sofa and was served good food, a tall glass of barley beer, or a clear birilé of mead, which often she left untouched.

Basha Deneqè's home was not far away. There, in the house she had stayed in before her husband's death, she felt more at home, more able to confide and, to an extent, to rest. She told the old commander where she had been, what she had seen, trying to make sense of it, to build a picture of the world in which she now moved. What did the arches say? for instance. They were proclaiming twenty-five years since Hailè Selassie's coronation; twenty-five years since he began remaking the city and the country, giving it an airline, a university, modern schools, hospitals, a theatre, factories, vast modern farms, usually named after himself or members of his family, and in the profits of which he often had a significant stake. Both of them also knew, could not but know, that these advances touched only a tiny fraction of the population, and further, that the emperor's five years' absence meant that from many he would never regain trust. He had gradually and skilfully achieved absolute power, however, and that, for years now, had been ample compensation.

But when the old commander's son visited she shrank back. It was not just his good looks and imposing size – though he was truly huge, as tall as anyone she had ever seen, and then taller again. She knew he had recently been promoted, becoming the emperor's personal aide de camp. Few could now approach the throne without going through him first. But she also saw that all this favour made the commander's son wary, that the gates of his house were barred against those who might use him as a route to power; she respected that, and did not want to seem to be asking anything of him.

As the months circled by it came increasingly often, unbidden, ambushing her when she was playing with her girls or chatting to her neighbour. When she was walking through the streets a smell could do it, or a glimpse, through a gate ajar, of oblivious women secure in their domestic round: a dart of homesickness so sharp that for a moment she could not breathe.

Or she dreamt, deep dreams, of a churchyard alive with bird-song and shifting shadow, and of herself in her spot near the south door, enveloped in sweet-smelling gloom. When she woke there was often a long moment before the stomach-thud of betrayal asserted itself. Or it arrived at once; and although she would shove the memories away again, scraps always remained. They coloured her hours.

Sometimes, feeling a little stronger, a touch more distant, she let the images linger, turned them over and over, thinking. She might not be on speaking terms with Ba'ata, but surely she was owed something? It was a wealthy place. Holy water was a thriving business, there were the acres and acres of land, the offerings due at festivals, funerals, baptisms, arbitration dues, market dues, taxes on craftsmen; it was one of the richest churches in all of Gondar. And she had seven children to feed, by the man who had rebuilt it. No – they had built it together: the mud and stone of its walls might be his proud bid for the afterlife, but they also contained her sweat, and her tears. Her life, even: ever since her wedding their histories had been bound about each other like vines. Why – the idea stole up on her – why not ask for it herself? Female aleqas were not unheard of. They could not actually administer, of course, a man had to be deputised to do that, but her brother was a priest, and entirely capable. Why not?

The next time she went to see Theophilos she bowed to the women. Bowed to him, and to his cross, closing her eyes, kissing it, looking for comfort in the metal tracery on her forehead and on her lips. Felt the nudge at her elbow, the burrowing into her palm, before she understood it, but when she stood straight she realised she was holding a soft fistful of banknotes. Theophilos's face gave nothing away, and neither, she hoped, did hers, even though her heart was full. He had been so good to her, so often. It wasn't the first time he had given her cash; and he had once arrived at her door followed by servants carrying bags of teff, and said, 'Let's eat together.' His gifts had seen her through weeks that in prospect had kept her awake, worrying about how she was going to feed her daughters. He had invited her to dinners where she sat mute, a slight, white-shawled figure surrounded by black-robed, black-hatted bishops. But when, one New Year's Day, she slaughtered a sheep and invited the neighbours, his chauffeur, his servants, she did not ask him. She had felt far too shy, that it would be too presumptuous. In later years she felt a hot regret; she real-ised that had she taken him food he would have appreciated it, that it would have been a small way of repaying him. Much later she tried to make up for her omission, sending the sweetest, finest chicken stew to him when he was imprisoned. He would not eat, had not eaten for days, but for her, came the report, he had tasted it and sent a messenger back with compliments.

She bowed, and answered all the questions put to her, about her children, the various priests in their mutual acquaintance, the latest twists in her quests at the palace and the courts, before she mentioned Ba'ata.

Not long afterwards she found herself in front of another monk, though this one, unlike his deputy Theophilos, was frail with illness and age. Basilios, leader of the Ethiopian Orthodox Church, listened. And assented.

She heard shouts before she saw anything, shouts and stuttering engines.

There seemed suddenly to be more people, some craning their necks, others being pushed back by figures she couldn't see, and she was caught in among them. Her nose filled with their various smells: dust and old sweat, rancid butter, bruised rue. They were too close. Too close!

Abet abet abet!

Ah. She knew what this was. A gap opened up in front of her, and she pushed forward. She was part of a wall of people now, facing another wall, in between them clear road.

The low, dark car turned into the space, creeping, almost silent, helmeted policemen running alongside.

Abet abet abet! Abet abet abet!

People began to reach out toward the vehicle. In each hand a piece of paper, held aloft like a flag. Choose me! Choose me!

The car slid through the paper forest, and slowed. Then the first person broke through the wall, dragged the shemma from around his shoulders, threw it down in front of the car. Another followed, and another, calling abet, abet, abet, calling in the way petitioners had called out to their emperors for centuries, around tents pitched on the northern plains and in the southern depressions, through royal enclosures at Ankober, at Qusquam, at Gondar, on hunting trips, at parades, at play or at war, stripping off clothing, knowing that the contract between sovereign and people was that no emperor could travel over garments laid down in his path.

The car stopped. The window was rolled down. There was that long nose again. There were children in the car. They must be his grandchildren. The people stared in, and the children stared back.

Abet abet abet! Abet abet abet!

The emperor reached out and took some of the pieces of paper. The window was wound up again and the car moved off. When she got back to where she was staying her daughters ran to meet her. Their bodies felt warm and solid. She buried her face in the hollows of their necks, breathing deep.

From across the courtyard her host the aleqa watched her. He nodded, but did not smile. Ever since she had announced that Basilios was giving her Ba'ata relations had been cool. Then a visitor had arrived from Gondar. He and the aleqa had talked for hours, and the climate had chilled further.

A few days later the aleqa came over to her little house. He had his own announcement to make. His visitor, appalled by Basilios's gift, had persuaded him – was he not more experienced, more learned? – to go to Basilios himself. And Basilios had reconsidered his position, and given Ba'ata to the aleqa instead.

The next time her son visited she sat him down and demanded he write out a petition.

*

ABET ABET ABET!!

The policemen crowded her, pushed her back, but she shrugged them away.

ABET ABET ABET!!

Day after day after day. And then one afternoon, 'Leave her be.' The busy hands dropped, and a space opened up about her.

Abet abet abet!

He stretched out a small hand, and she placed the petition in it. It was crumpled, browning in the creases. Then the window was rolled up.

The aleqa came to her again, and told her to leave. 'You see that I teach here. Deacons. Boys, and boys becoming men. It is not appropriate for you to be here.'

It hasn't been a problem in the past.

'You must go.'

Some time later she answered the door to a policeman. 'Madam, you are required to leave.'

I can't. I can't! I am the wife of a priest, I have made indissoluble vows. I'm still young, still fertile, I cannot move to ordinary lodgings, vulnerable to any passing man – and I have these children –

She saw the sympathy in his eyes, and the rising doubt, and it was not long before he left.

But the aleqa continued his campaign. So yet again she went to Theophilos, weeping this time, falling to his feet, and yet again he came to her aid. It transpired he owned a small house nearby, a bit closer to the slum, very modest, made of corrugated iron. He ordered it cleaned for her.

She spent one more day in the old compound, baking bread, decanting spices and grains, putting it all at the aleqa's disposal. And then she gathered up her children and left.

So they settled in to Theophilos's little house, the girls, their mother, and, at weekends, Edemariam, back from boarding school, who would pitch himself onto his chest on the low bed she had rented for him and lie there for hours, studying. When she was finished with the chores she took up her spinning and watched him, with a combination of pride and curiosity – the books looked nothing like the psalters and vellum commentaries that had so far meant education to her; many were, in fact, in a different language altogether. She worried about him, about the time he spent away, the amount of work he seemed to have to do. Not, as she might usually, about his health, or whether he was eating: even she could see there was no lack there, and that what they provided, at that school in the countryside six kilometres outside Addis, suited him well. Sometimes he brought crusty wheat bread home for them, light and rare, plain or spread with marmalade. She found that too sweet, in the way that she found honey too sweet, but the neighbours to whom she passed it on had no objection.

She was interested to learn the emperor took his role as minister for education so seriously he personally made sure the students ate well, visiting his schools regularly, distributing grapes, oranges, bread, cake. He had even been known, if a boy seemed especially peaky, to order food for him from his own kitchens. Sometimes he officiated at sports days: her son's school had a long swimming pool shaded by an old fig tree under which, it was said, Menelik II had rested on his way back from Adwa; boys who did well might find themselves bowing, dripping, in front of their emperor so medals could be slipped over their heads. The pool had new diving boards, Edemariam told her, as high as a two-storey house, and yes – enjoying how shocked she looked – he had jumped off them.

And all of this (bar underwear, pyjamas, shoes) was free, the only debt far in the future, nebulous but absolute: that each boy must use his education to serve his country, education being, the emperor had decided – why else was he minister of education, rather than minister of, say, finance, or defence? – the quickest way to gain on more developed nations. The expectation was so universal, however, so internalised, that for many it took on the flavour not of debt but of vocation. They began to look outward for inspiration, beyond their own country, through their foreign teachers in the first instance, at whom they sometimes laughed because they were so different from themselves but whom they more often emulated; she worried about the extent to which her son seemed to feel encouraged to ask direct questions, to argue, instead of doing what was expected of young people, which was to accept authority and precedent without question, that things were a certain way because that was how they had always been and always would be.

On Sundays she gave Edemariam fresh-pressed clothes and he went up the hill to a low, many-windowed building in the archbishop's compound for Sunday school. He accompanied his mother to Basha Deneqè's, or he went alone to the house of a new friend, a man she didn't know. Edemariam had been introduced to Germamé Neway by the civil servant who had watched Asratè Kassa rip up those first letters to the emperor; Germamé's father was the dean of St George's Cathedral and his brother led the Imperial Guard and so she trusted the connection, but from what she was able to gather from her son, the fastidious, traditionalist civil servant and this new friend could not be more different.

Germamé, a graduate of her son's school, had been studying in America until he was recalled by the emperor and placed in the ministry of the interior. He sounded kind (he gave her son pocket money, apart from anything else), bright, but argumentative and talkative. Idealistic. Her son returned from his weekend visits full of questions and assertions – about poverty, inequality, privilege; about the thousands upon thousands of gashas of land owned by absentee landlords and worked by peasants who owed so many tithes and taxes (the legal maximum being 75 per cent of the harvest, not counting inevitable bribes) they could barely survive. She could not help glancing quickly behind them, around them, at the corrugated-iron roof. Shhh! You must not talk like that!

He desisted, but was pleased to have been able to distract her. He hated her accounts of her days, of walking through the streets, buffeted by strangers, dependent on their erratic mercy. The waiting and the self-abasement. The sense of achievement, after she had handed the emperor her petition, had not lasted long. Many weeks had passed, and they had heard nothing, and so she had gone back to waiting for his car. It had stopped for her once again, but then there had been many more weeks, and still nothing. So she kept going to the palace, day after day. She returned to the princess, who had been sympathetic – so sympathetic, in fact, that, as she remembered it years afterwards, she offered her a house. 'Bring the rest of your children and live here.' The empress offered a place for Alemitu at her handicrafts school. Yetemegnu had bowed, thanking and thanking them, knowing all the while that she would not take up these offers, which served only to remind her (as if she needed reminding) how long she had been away – years, now. Getting on for three years. And she was so homesick for Gondar.

Edemariam hated her accounts of her days, but not as much as he and his sisters dreaded her nights. She hardly slept. She prayed, the Lord's Prayer, Gabriel's paeans to the Virgin, the Book of the Praise of Mary. 'She is the pillar of pearl that cannot be moved by the might of the winds, there is neither falling nor shaking for him that leaneth upon it.' Praying in the old language, the language of the church and of her youth. 'I know and I understand that I cannot comprehend it.' She had never used a rosary before but now beads clicked away the hours of the night, while her children lay curled into themselves, unable to sleep. Ayzosh, they would say to her, helpless, ayzosh. She replied politely, but she hardly heard.

So they cast about for entertainments. One day Edemariam took her into the front yard, broke a thorn out of a hedge and gestured, ney, come, sit here next to me. Then he leaned over and began to scratch shapes into the dust. A line swept up, bore left, then curved back in on itself like a noose. Yè, he said. She looked up at him and then down, suddenly and utterly focused on what he was doing. A plain cross, each arm of equal length. Tè. Then two ovals, like eggs, held upright by a straight walking stick balanced across their tops. Mè. A broken-backed vertical with a flat roof, and another small line, two-thirds of the way down, jutting out toward the right. Ngu. See? Try it. And she took the thorn from him and slowly wobbled out for herself the four letters that spelled her name. Yetemegnu. Those who believe. He smiled at her, a wide, pleased smile, and her face shone back.

Or they sat over coffee and asked her questions. What had she seen on her most recent travels? So she told them about priests trying to catch her eye when she went to church, about the messengers arriving with invitations to this party or that, the matchmaking matrons, the monk who knocked on the door, sat himself comfortably down, watched her moving about her house for a while, her youngest strapped to her back, and asked, 'Madam, is there no child who can bring us food, or pour mead for us?'

Who do you think you are? she had snapped back. Get out! She laughed, telling it.

Edemariam had always loved her laugh, which was low and full and entirely mischievous. Loved it more now that it was so rare. He watched her spinning, the practised snap of her wrist, and thought, she's still young. Her face is sad, but it's unlined, and she's beautiful. Why does she have to reject them all so completely?

'Why don't you marry again? We could have more brothers and sisters.'

What? Her answer was instant. Are you wanting a stepfather, then?

'Hmmm.' He thought. 'Maybe not.'

And that was that. But in the days afterwards she looked at him even more closely. How to tell him? How to tell him all that had gone through her mind before she had left for Debrè Libanos? How to tell him that she – and he, as the oldest son, however young and in need of a father he might feel – would head the family from now on?

How to tell him that finally she had a choice when it came to men, and she chose no?

Again she hardly remembered, later, how she got up the steps, how she crossed the hall, how long she waited, if she did, before she was ushered in; what the room looked like, who was there – apart from him, of course.

He was standing before his throne.

She bowed, deep.

He was king of kings, elect of God. But, as she had sometimes tried to remind herself over the long months, human too, with his own sadnesses. Le'ul-Ras Kassa, who had been at the emperor's side for forty years, advising, offering steadfast friendship, had died just a year before. Hailè Selassie had already lost three of his four daughters to childbirth and to illness, one of them in Italian captivity; a few months ago his favourite son, chosen, many said, for succession, had died in a car crash.

Your Majesty.

Your Majesty, my husband is dead. He cried out that he was burning, that he was being scorched from within.

No need to spell out who she thought was responsible – hadn't the emperor sent letters, after all?

He died a secret, terrible death.

No reply. Then a slight movement. She raised her eyes from the floor, toward the emperor's chest. It was his index finger that had moved, and was now pointing at the gilded ceiling.

'Tell that to Him.'

Eyes back down, instantly, to hide the gulfs of disappointment breaking open within her.

Yes, Jan Hoy.

'Is there anything else?'

My enemies have taken my land, they have taken my trees.

'That is a matter for the church authorities.'

Then, in a rush, in a bid to make him – a man, a father, the same age now as her husband would have been had he lived – to make him consider what her husband's death had been like, what it meant, what his enemies had been responsible for –

I have given birth to many children. They are scattered across his ashes.

'They shall go to boarding school.'

Yes, Jan Hoy.

And then she was being ushered out, bowing, being handed – what? A few birr, money 'for the journey'. What journey? Laughs too loud, remembering it years later. What was it? A bribe? Charity? Insulting, anyway.

Those were bleak days. The urge for justice drove her still, but now it had nowhere to go. Should she have spoken differently? More persuasively? No, she had said what she needed to say. And he had understood. Perhaps he had even sympathised, but what use was that? What a burden for her children. What a burden for her. What a waste.

As for the schooling – they were too young. The smallest girls especially, they were far too young. She wanted desperately to keep them with her, to feed them well, to hold them close. Alemitu was old enough, of course, but her case was different. She was deaf, and therefore vulnerable, especially in a big city. Who would watch out for her, make sure no advantage was taken? She could sell the rental rooms, she would do whatever was necessary, but she could not send them away. They needed her. And they were her companions and her closest friends, now more than ever.

It took some time, well after she had told people what had transpired, after they had responded with varying degrees of envy, depending on how many children they had and what kind of schooling they were receiving, if any, before she realised she had no paperwork, no proof of the emperor's order. So she made her way to the ministry of education. But they were practised at turning people with this particular request away. Jan Hoy has directed you to educate my children. 'Do you have proof?' No proof. 'Be patient, madam. Be patient. Wait.'

After four months of waiting she went to see Basha Deneqè's son. His gate was shut, the walls were high and barbed. But she stood there until he came out, and threw herself before him, begging that he do this one thing for her. Please, upon your father's death, please bring me to Jan Hoy.

She felt him considering her from his great dark height.

'Ayzosh,' he said eventually, raising her to her feet. 'Ayzosh. I will bring you to him.'

On the sixteenth day of the sixth month, which is the feast of the Covenant of Mercy, she came before the emperor one last time. Hailè Selassie repeated his order; the court recorder, this time, did his part.

As soon as she had pushed the letter over a counter at the ministry, after it had been stamped and, miraculously, ratified that same day, she packed up her little house by the cathedral and took her daughters home.

BOOK V
1959–1989

YEKATIT
THE SIXTH MONTH

Dry. East winds bring puffy clouds and hungry locusts. Some ripening of wild fruits. Late barley matures and is harvested. The price of meat doubles.

AND AGAIN THE ANGELS SET HER SO THAT SHE MIGHT SEE THE PLACE OF JUDGEMENT ... AND OUR LADY MARY SAID, 'WOE IS ME! WHO WILL ANNOUNCE TO THE CHILDREN OF MEN THAT THEY SHALL COME HERE?' ... THEN THE ANGELS CARRIED HER ALONG AND BROUGHT HER BACK TO HER FORMER PLACE. NOW THAT DAY WAS THE SIXTEENTH DAY OF THE MONTH OF YEKATIT, AND SHE STOOD UP ON 'THE PLACE OF THE SKULL' (GOLGOTHA), AND SHE MADE SUPPLICATION UNTO HER SON ... AND JESUS CHRIST CAME DOWN, AND WITH HIM AND ROUND ABOUT HIM WERE THOUSANDS AND TENS OF THOUSANDS OF ANGELS. AND HE SAID UNTO HER, 'WHAT SHALL I DO FOR THEE, O MARY MY MOTHER?' ... AND OUR LADY, THE HOLY VIRGIN MARY, MADE ANSWER ... 'WHOSOEVER SHALL CELEBRATE THE FESTIVAL OF MY COMMEMORATION, OR SHALL BUILD A CHURCH IN

MY NAME, OR SHALL CLOTHE THE NAKED, OR SHALL
VISIT THE SICK, OR SHALL FEED THE HUNGRY, ...
REWARD THOU HIM, O LORD, WITH A GOOD REWARD
FROM THYSELF.' ... AND OUR LORD JESUS CHRIST
ANSWERED AND SAID UNTO HER, 'IT SHALL BE EVEN AS
THOU SAYEST, AND I WILL FULFIL FOR THEE ALL THY
PETITION. DID I NOT BECOME MAN THROUGH THEE? I
SWEAR BY MYSELF THAT I WILL NOT BREAK MY
COVENANT WITH THEE.'

— *THE COVENANT OF MERCY*

It seemed to her that she was walking down a familiar street. Under her feet familiar stones, over her head familiar sky. The dust of high summer caught in her throat.

When she looked up, a woman was standing under the branches of a wild olive tree. Its narrow leaves cast a filigree of shadow over white shawls and a pale face.

Her heart leapt. But she was also wary, and hung back.

The woman came toward her, arms outstretched. 'Come, child, kiss me. What, have you forgotten me?'

Years of hurt burst from her then. Oh, but it is you who has forgotten me! I would have come to kiss you, to bow to your walls, to pray to you, but you abandoned me!

'No, child,' replied the woman. 'I have always been here. It was you who abandoned me.'

*

For the first year both the girls and the boys attended schools in Gondar. That she was able to countenance. But then her half-brother Gebrè-Selassie came to tell her there weren't enough pupils to sustain the girls' school, and her daughters would have to go to Harar – Harar? Harar? Her voice rose. A Muslim city, far to the south-east. An ancient city, thronged with chat-chewers by day, was all she had heard, its walls assailed by hyenas at night. Above all, so far away.

But they are so small! They cannot go!

They must go. You've been so strong, such a man, will you be weak now? The girls will be well raised, by the most modern methods. And there will be nurses for the youngest. They must go.

Let them stay! I'll sell every room, every tree –

After you went through all that pain? For all those years? After God himself came to your aid? I forbid it. They must go.

Eventually he relented about the smallest. But when the rains ended and the hills were clothed in golden flowers, she took the two middle girls down the long road to the airport and saw them off on a plane to Addis Ababa. Her son would meet them and put them on the train to Diré Dawa, and from there they would travel in a car to Harar. Years later Tiruworq would remember that when they arrived at the school they were greeted with amazement. What were they doing here, so young? Had their mother died?

For days she could not eat or sleep. They're going to be so accomplished, they told her, attempting comfort. So modern, so educated. But she would only turn away and weep. Her youngest watched her overtly, watched her covertly, whispered to her in the middle of the night, when the pillows they shared were soaked and it seemed as though the invocations to the saints would never end, Ayzosh, Nannyé. Ayzosh. Yibejish, lijé. Take heart, mother. Yes, child. May you be saved. But it made no difference.

Relations with the eldest were more fraught. The husband they had chosen for her drank to excess, it transpired; while she was away Alemitu had discovered he was going through her father's library, taking hand-illuminated church manuscripts to sell them for mead, and kicked him out. Now she cleaned and cooked and pounded spices, she spun cotton and embroidered and wove baskets, she cared for her brothers and sisters and watched as they were readied for a future she knew she was capable of but could not have, and when the resentment and frustration rose too high in her screamed at her mother that she was being left behind, nothing was being done for her, why was nothing being done for her? But what do you want us to do? Yetemegnu would wail in return, facing Alemitu directly so her daughter could read her lips. I am protecting you by keeping you at home. If you were well you would not speak to me like this. But Alemitu knew that her mother had turned down the empress's offer and was not appeased.

When her sons came through the door on weekends or high holidays, all slim dark limbs and shining eyes, Yetemegnu's heart would lift and she would fuss and fuss, making fresh injera dripping with butter and berberé, sitting them down, asking them about every corner of their new lives. Nannyé, we get up at six, said Molla, and there is marmalata for breakfast, and foreign butter, and tea and milk and bread and eggs, for lunch there is macaroni, sometimes soup. The teachers are Indian, they know everything. The classroom walls are covered in useful sayings and proverbs. After class we go to the study room, and everyone is so quiet, reading – Sometimes we go into the garden to water our plants, interjected Teklé, who had always loved to do this, and there is a male matron to look after us. And everyone, rich and poor, has to wear blue khaki. Princess Hirut said so.

She searched their faces and tried to imagine their new lives, and loss was entangled with pride and hope and became indistinguishable. They saw this and were silent about the journeys through the city that bracketed their time at school, how each Saturday they would break away from the shortest, most direct route home and weave through stony back streets to avoid the children who pointed, whispered behind their hands, or jeered. They had always known, in their father's last years especially, that they were not allowed to accept ad hoc hospitality; now, angry and humiliated, they did not want to. They had become wary even of children with whom they had grown up. To the cold sense of their father's absence – exacerbated on holidays such as Masqal, when schoolmates in fresh white jodhpurs and bright new shemmas followed their fathers to the bonfire below the castle walls – was added a sense of being fundamentally alone. Much later, Molla in particular would wonder if their father was not in fact ahead of his time, a kind of revolutionary, even, but now there was nothing but home and school, and they understood, with a clarity so pure it was almost a physical pain, that education, for them, would be all. They knew that if they had told her about their walks home she would have understood, and enveloped them in sympathy. But also they knew how hard she would take it, and so a habit of silence began where their mother was concerned; a sense of care, and a duty not to worry her – she was worrying about enough already, worrying too much.

After her dream she had started to go to Ba'ata again, stepping through the streets, eyes fixed on the hard ground, shemma over her head and tight about her shoulders. The women parted as they always had, allowing her back into her place, but when they poured out of the church after the service she felt fewer of them greeted her, that the welcome was not as warm as it had once been. She looked around her, at the trees, the worn outbuildings, the piles of hewn stone still lying under the trees.

The threshold at home, she would insist to anyone who would listen, was growing grass. No one stepped over it. Where were his students, who once had eaten her food so enthusiastically, who had drunk her mead and sung her praises? The deacons and priests he had supported and promoted? Though in fact some people still did drop by and sit for a while, sipping barley beer. Gaping holes would open up in the conversation, which flapped and lagged like a web ripped through by a spider's prey. Or it was too garrulous, overloud. Again and again she found herself saying the same things. Harsh things, laced with incipient tears. They ensnared him. They killed him. They wanted to get him for years and finally they succeeded. It was like a compulsion she could not control. She looked at her neighbours and friends and could see or think of little else. They ensnared him. They hated him. They killed him. They are after me now, and after my land. Her visitors sat in front of her murmuring awkward assent, and did not always come back.

But one visitor did stay – a scrawny son of her favourite half-brother's who late one night appeared panting at her door, having run away from home after one too many arguments with his stepmother. May I stay? Please? She looked at Alemitu, for whom it would mean yet more work, and Alemitu looked back at her. The next morning the boy, Alemantè, went to school, and she went to his stepmother, to ask for his blanket.

They set about knocking the cobwebs out of the house. The one-room properties that flanked the entrance to the compound, fifteen in all, had never been empty – a couple were rented to poor householders, some went to sellers of barley beer, others to market traders from Asmara, who piled them high with bars of salt, bolts of brilliant-coloured cloth, sacks of teff, great brass measuring scales, and waited, vigilant, for the prices in the market to peak. Her husband, far-sighted, had circled the main house with more rooms, each of which had an entrance into the yard. Alemitu took two; various relatives moved in and out of the others.

For herself she took charge of the big central room, dividing it with curtains, making beds in the far corners, leaving the centre as a living room, where, with great ceremony, she installed a radio. It was a huge radio, encased in wood, protected with embroidered runners. Three times a day, every day, she asked for silence, or, if it was not immediately forthcoming, enforced it with a sharp 'ish!' then sat down to listen to the news. She listened for mentions of Gondar and of Addis Ababa, for the names of those she knew or had met, and of those, thanks to this amazing box, she felt she was coming to know. She drank in accounts of events across the rest of the country, and in the world beyond its borders, trying out on her tongue the names of foreign leaders, entering, over the following decades, wholehearted into their achievements and their anxieties, weeping for Samora Machel, listening intently to the progress of Indira Gandhi – ayi! If I'd gone to school I could have been like her! – absorbing all she could about the myriad countries in which her children and her children's children came to live.

When the girls came back for the rainy season, held together by a short rope so they would not lose each other, reprising the stages by which they had travelled – Harar to Diré Dawa by car, overnight train to Addis, where they were met by Edemariam (whose holidays were not paid for by the government, and were therefore less frequently spent at home), by plane from Addis to Gondar – her ililta, her hugs, her kisses were planted deep in the hollows of their necks and punctuated by hiccuping tears. She was so happy. Come, come, let me look at you. Kissing and kissing them, holding them tight, as though she would never again let them go.

Then at last, come, come, look what I have for you. And she would lead them through to where the food was waiting, buttery porridge and fresh injera and dish after dish from a just-slaughtered sheep, meat in hot sauce and in mild sauce, pan-fried ribs, spiced offal. The room would be full of people, too, relatives who had come from Simano, from Atakilt, from all over Dembiya, bringing their own offerings, some of the farmers so poor that even the girls, young as they were, knew these dignified meek-seeming men could not afford the battered tins of beer, the baskets of muddy eggs nestled in straw they bore so carefully into Gondar.

And she would hover about them, Eat, eat, upon my death, please eat, here, here is the best bit in the whole dish, you are not eating, what is wrong, please eat. Searching their faces, Tell me about Harar. Do they feed you? What do they feed you? Do they care for you? Yes, yes, they are good to us. They tell me there are hyenas, I have been terrified, wiy! I have been terrified that the hyenas will take you. Are there hyenas? Yes, there are hyenas, but they're kept outside the walls, they don't often come into the city, Nannyé. In the mornings she brought them breakfast in bed, injera still warm from the griddle, dripping with butter and spice; when they got up she gave them new clothes to wear; in the evenings, after supper, she would gather them around her. Terèt, terèt! And they would reply, delighted, yelam berèt! There once was a jester called Shiguté. One day he went to his master, who was a grand military commander ... and they would settle at her feet, stilled with the pleasure of attention. Sometimes they sang, competing to remember the wittiest verse, or the most moving, and sometimes someone would turn the dial on the radio and they would find themselves standing, shoulders shaking, heads dipping forward, back, forward, back, dancing and laughing, surprised by joy.

There is a rare photograph of them all, plus various half-brothers, uncles and cousins, taken about two weeks into the new year, just before the annual dispersal to school, and, for Edemariam, college. In five days it will be Masqal, when the tabots come out of the churches and process around high bonfires lit in celebration of the arrival of a fragment of the True Cross in Ethiopia. But the holiday also coincides with the end of months of heavy rain, when the sun reappears and the fields, wet and lush with barley and masqal flowers, reflect its rays back into the bright, humming air. In fact, the sun is out now: it picks out each blade of grass in the yard and outlines every stone in the walls of the house. It beats, fresh and hot, into faces accustomed to days and weeks of dark and damp indoors, revealing every contour of forehead, cheekbone, nose, mouth, chin, bleaching the cotton of shemmas, of best dresses, a brilliant undifferentiated white. Nearly everyone squints into it, or gazes up from lowered brows, or looks to the ground; one small boy gives up altogether and shuts his eyes. Alemitu, a defiant Nefertiti, faces it directly down.

They are almost all entirely serious, as was normal in photos then, but two manage slight smiles – a girl cousin, and Edemariam, tall in the back row, arms draped protectively around the shoulders of Tiruworq and of Teklé, who despite being older is not much taller than her. Although Teklé, along with his father, had recovered from the fevers that killed Yohannes during the Italian war, he had never really thrived, eating little, physically careful, retiring and kind and watchful except for the occasional rage. And then he had contracted TB, and scrofula. An abscess appeared on his neck and he became ill again, shaking with chills, losing even more weight. Again she did the rounds of the churches, looking for the holy water that might cure him – Loza-mariam, near Azezo, Seqlè-mariam, two days' mule journey away. A traditional doctor gave her sulphur, which she hid, taking out tiny amounts to drop onto the abscess and shrink it, but Teklé, impatient, found it and dosed himself with too much. It burned through the abscess but also through great sections of his neck, scarring it and altering his voice. At last she took him to a zar-doctor, who prophesied a long life. The abscess did not recur, but he had missed a great deal of school and would never really catch up. She always said the photograph was a sad occasion because her husband was not there to take his place at the centre, but she is the only one who smiles a full, open smile.

So she was home at last, but returning had not freed her from some of the more onerous routines of her exile. She was still fighting for the land at Bisnit, for instance, where the eucalyptus trees reached higher and higher and the monks, despite their supposed renunciation of worldly values, grew ever more intractable. At court she argued that she had never denied part of the land was theirs, of course they could grow vegetables along the banks of the stream, of course they could have that, however it seemed to her they wanted it all now, land, trees, crops, and that she could not allow. There was another skirmish too, with a family that had taken advantage of her absence to annex some of her rental rooms.

On court mornings she woke jangling with apprehension. Her children would watch her dress – an underdress, a plain dress. Then another dress and another on top of that, because she was so thin, and did not believe she could be taken seriously if she was not more physically substantial. One netela, another. And with every layer, a litany, why this again, why this, look what I have to do, I'm only doing this for others, look what I have to do. So she worried aloud, and prayed, until finally she left the house in the company of whichever male relative was free that day – her half-brother, perhaps, or the boy Alemantè, who came with her when there was no school and soon became adept at writing memoranda replete with lawyerly phrases he did not always quite understand. When she returned she did so cursing the courts and everyone in them, the days and everything encountered during them.

Other times a child, Alemantè or Maré usually, would walk into the main room and find her sitting empty-eyed on the daybed. They knew what generally came next, and they hated it. And sure enough, it would begin – head first, tipping to the side, dropping forward, tipping, dropping back, again, again, wider, and wider, each ellipsis bringing more of her body with it, then a voice whipped into the air like seeds sown onto a stiff breeze, a voice they did not recognise, full of words, some of which they understood, many of which made no sense to them at all.

Weliyé, Weliyé, Weliyé,
Weliyé's axe has two mouths,
You don't see him cutting till the tree falls.

At last she would slip into a trance and then into sleep; finally they could creep out into fresh air, and breathe.

But for all the evident and much-performed anguish they knew it was also true that all those years in Addis meant she was now a practised litigant, equipped with ploys and defences honed on a far larger stage than this. So she was not unduly impressed by the dirty-yellow Italian courthouse at the centre of town, whose high arched windows looked unfeeling down at the ranked shacks of scribes writing letters and testaments for the illiterate. Here, on home turf, she knew who to nobble and how, where to stand, which mornings it was worth going and which it wasn't. She knew how to go through the papers she was given and commit to memory a thumbprint or a smudge, the sweep of a signature or a torn or fading corner, so that even though she could not read she could select with confidence the correct documents when required, wrongfooting men – for the judges were always men – who assumed the rare women who came before them had no such ability.

She learned to do the same to the inevitable suitors, too, choosing circuitous paths home, affecting incomprehension, even, on one occasion, picking up her skirts and bolting. She had accepted a dinner invitation from a nun of her acquaintance and asked Alemantè to come with her through the dark streets. He was sitting in an anteroom dozing over his schoolbooks when she reappeared, whispering and shaking him, Come, come, we have to go, we have to go. He leapt up unquestioning, and then they were both running down the hill, trusting their feet to find safe grips among the cobblestones, running past the castle compound, breath loud in their ears. When they got home she said it had suddenly become clear the dinner was an enabling of a male guest's interest in her, and that when he had made his move she had simply stood and walked out. Thank you, child, thank you, she said then, and for years afterwards, when increasingly laughter mingled with the telling, thank you for getting to your feet so fast.

And then one morning she arrived at court with her half-brother and a witness, a priest who had agreed to support her suit. Proceedings began as usual, but when the priest began to speak it became clear he had at the last minute decided to switch sides.

Erè, erè! she said before she thought.

'Quiet,' said the judge.

'I am telling the truth,' insisted the priest. 'Our Lady knows I am telling the truth.'

Yes, and Our Lady also knows how you pray, she shot back. She has watched you bowing in the light of the lamp. Everyone understood her; he had been having an affair with a woman whose name meant 'light'. The room rang with delighted laughter, and the priest subsided protesting into silence.

'Aleqa Tsega's gifts haven't died,' people began to say, giving her her backhanded due. 'They've simply been passed on to his wife.' But when she entered the chambers each morning, she did so to respect that was hers alone.

The issue of the houses was soon decided in her favour. The case for Bisnit, however, would take another decade.

MONDAY

She still celebrated Ba'ata each year, though not on the same scale as before. She brewed beer, but only five gans rather than thirty or forty. There were a few stews, and a pile of injera, for priests and for family who had walked in from Dembiya and Infiraz and

Gonderoch Mariam to be with her. For the students and lower-rank church people who would go anywhere for free food she roasted a quintal of dried peas.

The students were all very well, she often liked them, and she imagined that their mothers, far away in other provinces, would be pleased to know they were being fed, but the priests and the priestlings – she smiled and greeted them, asked after their families, played the kindest host, while behind her eyes the commentary circled and jabbed and threatened to break through. Where were you when he needed you? Did you contribute to his trouble? And will you now go back to your dark huts and gossip about how Aleqa Tsega's family has fallen?

And behind the familiar, curdled litany, other, more complex feelings. Fewer guests meant less work, surely she should be grateful. So why this itchy, achy, private sense of loss? Mother, said neighbours who noticed her abstraction, this is so good, greater than how it was before. The beer hits its mark. It wasn't about the death of her husband, exactly, though that was the cause of it. What she missed was the camaraderie and the gossip, the flamboyant complaint and the laughter, the shared purpose of dozens of working women pulling together. To her specifications, of course; a grown-up echo of the games of her childhood, when the other children pelted through the fields and thorn-hedges but always came back to her sitting collected on her mound of rocks. As a landowning widow and mother, well-born, she was now granted more authority and autonomy than most women would ever possess. She knew the value of it. But at times like this she found it was not quite enough. Where were the rases, the dejaz-matches, the bishops? Where was the sense of being at the centre of things? She said none of this, of course, just greeted her guests warmly, smiling, bowing, bowing.

TUESDAY

The next day the crowd was thinner, composed more of relatives than of churchmen and nuns. Moving among them, persuading them to eat more, and more – upon your father's death, upon my death, but you must! No, not that poor scrap, don't be silly. Here, here's an especially tasty bit, let me feed you, no, no, open your mouth – she felt more at ease, more able to give herself up to chat, to the play of anecdote and memory, to her glee in wit so sharp the victim took a moment to notice a mortal wounding.

WEDNESDAY

After the first and the second and the third rounds of coffee, after the frankincense had melted into the coals and filled the room with banking clouds of sweet smoke, after her benediction, said in a monotone, hands before her, palms to the heavens – May He bring justice to the wronged, and to the poor and the oppressed. May He clothe the naked and liberate the crucified. May He protect us from sudden death, and bless this house with longevity and with plenty. May He who brought you safe from all your lands return you there in safety – her guests replied, Amen, Amen, Amen and collected up their shawls and walking sticks and loaded their animals and kissed her on each cheek, one, two, three, four, five times and bowed through the gate and away. When they had gone she turned back into the house, loneliness and worry momentarily at bay, and asked the servant girl to sweep up the grass and attend to lunch.

Back in the main room she reached for the radio and switched it on.

… Just as the bee whose hive has been disturbed becomes violent so also have some opponents abandoned all restraint. But despite that they have been unable to hold back a growing awareness among the Ethiopian people.

The individual words made sense, in and of themselves, but …

The few selfish persons who fight merely for their own interests and for personal power, who are obstacles to progress and who, like a cancer, impede the nation's development, are now replaced. And I have, as of today, agreed to serve you and Ethiopia as a salaried official under the Constitution.

I? Who is this?

Know that all decisions and appointments declared by the new Ethiopian government formed by me, and supported by the armed forces, the police, the younger educated Ethiopians and by the whole Ethiopian people, are effective from this moment on.

What new Ethiopian government? Where is the emperor? What is going on?

People of Ethiopia! Let your unity be stronger than iron bonds! Today is the beginning of a new era for Ethiopia in the eyes of the world.

She sent a servant chasing to a neighbour's to place a phone call, but the lines were silent. There was no way to reach Addis.

There was only the radio. She sat in front of it, willing it to explain. 'I', it transpired, was the crown prince. The emperor was abroad. But the younger educated Ethiopians, the younger educated Ethiopians – what was her son doing now? Where was he? The radio had no answer to that question. When finally she peeled herself away from it to go to bed she prayed and prayed and could not sleep.

THURSDAY

It was not yet light when she dragged herself aching out of her bed and into the cold back yard for her ablutions. Eucalyptus leaves rattled faintly, silver-green sabres. A cock crowed, harsh in the stillness.

Coffee, and the radio again. The aims of the new government, 'peacefully established in the Empire under the direction and leadership of the crown prince', with Ras Imru as prime minister. Ras Imru? How interesting. She had known he was critical, progressive even. Young people currently shut out of education would, it was promised, be welcomed back in. There would be new factories, and much help given to farmers.

A group of bandits under the influence of two traitors – former generals Merid Mengesha and Kebedè Gebré – opened fire on peaceful civilians demonstrating for the new Representative People's Government. Several of the civilians were killed in this inhuman massacre and many were wounded –

He's dead! He's dead! My son is dead! Howling, undone. He's dead! He's dead!

Neighbours, brought running, 'Ayzosh, you don't know for sure' – He's dead! He's dead!

'Hush,' said one of her neighbours. 'HUSH!'

Urgent announcement. Urgent announcement.
All people must evacuate themselves as soon
as possible from the area near the railway station,
near Mitchell Cotts commercial offices, Menelik
Square, Mesfin Harer District and the
airport.

They brought her a sash, and helped her fumbling hands tie it tight about her waist, to hold the pieces of her together.

On the radio one voice claimed imminent ceasefire, another claimed increasing support for the new government. Troops were stationed across Addis. A curfew was declared throughout the empire.

She curled up on the floor, hugging herself, rocking. He's dead. He's dead. He's dead.

FRIDAY

The anniversary of her husband's death. Seven years. Each year she cooked – or had cooked, if she was in Addis – split-pea sauce in a large pot, exactly forty-two fresh injera, and a gan of tella, and had all of it carried up the rise to the church, where priests from Ba'ata and often from other Gondar churches stood at her husband's grave and prayed for his soul. After communion they gathered near the eastern entrance to the church, the so-called gateway of peace, and there they ate the food she provided in required payment for their pains. There was so much that often they kept some, and ate it for a second day.

The anniversary both compounded and mocked her present grief. She sat unresponsive among household and neighbours, and the radio, to which they all now felt umbilically attached, talked on and on, one voice replacing another, until, finally, an announcement that the revolt had failed and the emperor was home. The army were on their way to free royals held captive in Genetè Le'ul.

The memory of the emperor's palace came to her so clearly she could almost feel it: the tall narrow windows, the columned portico, the palms, the steps climbing into nowhere – They were reading a list. Le'ul-Ras Seyoum Mengesha, grandson of Emperor Yohannes IV, and governor of Tigré, dead. Ras Abebè Aregai, chairman of the Council of Ministers and minister of defence, dead. Mekonnen Habtè-Wold, minister of commerce and industry, brother of the deputy prime minister, dead ... on it went, fifteen dead, many of high rank, shot at close quarters in a locked room, blood sprayed across carpets and wooden floorboards, high leather-backed chairs, pale pillars. Brigadier-General Mekonnen Deneqè, hit by many bullets, alive, but badly wounded. She imagined the huge body, felled and suffering, and could not understand when the tears were going to stop.

The rebels had written, they said, the 'first honourable chapter in Ethiopia's history, which will be remembered forever', even though it had failed. That if it had succeeded, it would have freed the Ethiopian people from poverty, ignorance, slavery. That the coup leaders 'did not value their own honour and wealth but were only instigated by seeing our problems and suffering', and should be avenged. That 'our fortunate and happy brothers, Mengistu and Germamé' – Ah, her son's new friend. It was a piece of the puzzle that slotted into place so easily it was as if it was already there.

The moment communications reopened she began searching for Edemariam. Oh, the wonder of his voice! I'm fine, mother, please don't cry, I'm fine. Relief and tears, anger and tears, when she discovered the massacre was an invention meant to turn people against the generals; no one had in fact shot at demonstrators. There had been fierce fighting, however, and bombing from the air, and many civilians had died before the ceasefire was finally announced. And yes, her son had marched against the emperor with all the other students, and after the march had been broken up and they had returned to their dormitories he had been craning out of a window to see what was going on when a bullet whipped past his ear. So her instincts had not been entirely amiss.

For weeks the country thrummed with stories, patched together from unofficial eyewitness accounts and official statements in attempts to produce a semi-comprehensible whole. She heard of blood flowing through familiar streets, of bodies toppling off Ginfillé Bridge into the brown water below, of days and nights of gunfire. She heard of a bounty placed on the heads of the Neway brothers and of their capture on Mount Zuqwala, the holy volcano just south of Addis Ababa. Some said it was Germamé who shot his brother and killed himself; others that Mengistu shot Germamé. Both stories caused her to raise her hands to the heavens and pray for their souls. That one side of Mengistu's handsome face was blasted through, but he lived and was taken to hospital. That the emperor ordered Germamé's body strung up in front of St George's Cathedral.

For days the emperor's level voice scraped through her living room. 'You all know how much We trusted and how much authority We reposed in those few who have risen against Us. We educated them. We gave them authority. We did this in order that they might improve the education, the health and the standard of living of Our people. We confided to them the implementation of some of the many plans We have formulated for the advancement of Our nation. And now Our trust has been betrayed ... The judgement of God is upon them; wherever they go, they will never escape it.'

Prison camps were built and filled. Students were pardoned (though many signed letters accepting any future punishment for a 'traitorous act against a parent who has brought us up, who is our father and our mother'). Asratè Kassa, who had organised much of the counter-attack on behalf of the crown, was promoted to president of the Senate, while his nephew became governor-general of Begemdir. Brigadier-General Mekonnen Deneqè became minister of state in charge of security. Newspapers indulged in elaborate puns, reported gleefully by those who could read to those who could not, and underground leaflets were everywhere. 'What is sinful is to be ruled by despots, not to rise against them!' proclaimed one. 'If the branches of the forest trees united they could trap the lion,' argued another.

A few months after the attempted coup, Mengistu Neway was brought to trial. He had not intended bloodshed, he said. Leaders were meant, as the emperor himself had claimed, to 'be servants of the people, not vice versa', but few remembered that. It was time for a fairer Ethiopia. Throughout the seven-week trial the city was tense, the army restive and demanding higher pay, which, from a shaky-voiced emperor, they got. After seven weeks Mengistu was, to no one's surprise, convicted, taken to the main market, and hanged.

*

After the rainy season she travelled to Addis. Abunè Theophilos, now deputy to Ethiopia's first independent patriarch, had written to the World Council of Churches on her son's behalf, and it had responded with a scholarship. And Edemariam, who had never forgotten watching his brother die, had chosen to study medicine.

She knew she should be happy, she was happy, but she was also desolate. She sat and watched as her son prepared himself. She fussed and fidgeted. She could send him off with food, but of course it would not last. What could he possibly eat, in a place called Canada? He would fade away, lonely, among strangers.

I had a dream, she told him. I saw a mark on your forehead. It means that in the first country you go to there will be someone to meet you. Someone who will help you, like your mother or your father. He smiled at her, but they both had to know this was a reach, a brave flailing for comfort. Not long after he arrived in Montreal he sent her a photograph of himself standing at a stove, stirring something in a pot, as reassurance; she composed a poem saying that at last she could sleep. In fact, he never did learn to cook.

The emperor no longer had an audience with each student who went abroad – there were too many of them, and since the failed coup he had removed himself somewhat from day-to-day involvement in education. But he had been at her son's graduation, handing out diplomas, and his message, as always, was clear: go, study well, then return to serve your country. The irony in this was now lost on nobody: it was those he had sent abroad who brought back ideas of governance that ran so counter to his own; or, as the emperor himself put it, 'Trees that are planted do not always bear the desired fruit.'

The Jesuits who ran the university had given Edemariam long

grey trousers and a blue blazer, the emperor had provided a knee-length coat, and one day just before he left he donned his finery and they all went to Trinity Cathedral. The sun, testing its rays after months of rain, warmed the dark stone and burned through their clothes so after the service Edemariam took off his coat and laid it carefully over a fence while he talked to his mother, his sisters, his brother. He noticed a boy loitering, but there were always boys loitering, and he thought nothing of it. When he turned back for his coat, however, it was gone. At once he was running – and so was she, dignified widow though she was meant to be, picking up her skirts and chasing round the fence, up the steps, into the church, collaring the gibbering child whose attempt at innocence collapsed at the first slap. 'It's in the crypt! It's in the crypt!' Edemariam retrieved it from under a pile of dirty blankets, laughing and shaking with relief at a bad omen averted.

They said goodbye to the archbishop, and to the brigadier-general, who beckoned her son over and handed him more cash than he had ever handled in his life. She organised a small party, her children, a few relatives, neighbours. It was meant to be a joyful send-off, but her grief was catching. Guest after guest turned away to hide eyes glittering with tears. Everyone tried to bolster her, saying how brave she was, how lucky; the archbishop told her off: 'He's going to study. Do you know how many people have not had his or your good fortune? You must stop!' but she knew now how this kind of bravery and this kind of luck actually felt and she could not be persuaded. The day before he left she sang to her son a verse she would sing often over the coming years:

Parting is death
For those who love,
Parting is death
For those who love
To be buried standing.

She was startled when he joined in. He was not a singer – it was
the first time she had heard him sing anything apart from childhood ditties and scraps of liturgy, and it would be the last. She
hugged him as if she would strangle him, while he tried to comfort
her. Ayzosh, Nannyé. Ayzosh. When she went to see him off at the
airport, she felt she might dissolve with sorrow. What if she never
saw him again?

Every Friday she washed, standing in a wide metal tub while a
servant girl or granddaughter poured jugs of water over her then
rubbed her down, scrubbing her back, her arms, her thighs, until
every curve and plane of skin offered itself up tingling to the air.
She lifted clean white clothes over her head, embroidered white
muslin, a matching netela, a headscarf. She drank coffee, as on
every other day, except that on Fridays she took more time over
the ceremony – cut grass laid down, her best carpet, the best
incense, making sure every step, from roasting to pounding to the
three cycles of pouring and drinking, was properly, unhurriedly
honoured. Then, not every Friday, but often, and especially on
certain feasts of the archangel Mikael, she would go to the chest
in her bedroom and lift out strings of coloured beads, red, green,
yellow, blue; long necklaces of black silk wound in gold and
ending in a wide gold ring; bracelets to climb her wrists and traceries of silver and gold to rest on her ankles; she threaded rings

onto her fingers, then, spraying it on the insides of her wrists first, brought to her neck, to the front of her dress, touches of Sudanese perfume.

When dusk began to gather the houses in, huddling them under a sky brighter and somehow higher in these moments just after the sun had left; when the shadowed hills were crowned with gold and kites owned the heavens; when the streets filled with a sense of focused hurry, of errands to finish before the light was switched off altogether; when the air thickened with cooking smoke and a day's worth of kicked-up dust, she walked out of the compound and, accompanied by a daughter or a granddaughter or by Alemantè, turned right toward the market, then left toward the mosque and the Muslim quarter. From a low thatched house on a narrow lane came the sound of voices, and the intermittent, exploratory thump of a drum. She knocked and was admitted. The children were welcome, and the girls would often stay for a while, but Alemantè always declined.

The room was thick with incense, the light from a few small lamps struggling against the haze, the gloom full of people, mostly women, sitting on the floor. They smelled of perfume too, the scents competed with each other; around their necks and arms and ankles hung beads and chains like her own. She handed the offering she had brought to an attendant – a pitcher of beer, or some food, or money, and bent to take off her shoes. Then in bare feet she walked over to the far end of the room, where a curtain was drawn across a raised-earth dais. She kneeled, and kissed the ground before it.

Greetings.

'Welcome, Yetemegnu,' came a voice from behind the curtain. 'Are you well?'

I am worried, madam, worried about my son, I have not had a letter for months. Or, my younger son is ill again. Will he ever be well? Or, my daughter is unhappy, I do not know how to help her. Or, I have been ill myself, madam. My heart aches. I am afraid, always, I worry and I cannot sleep.

Sometimes the reply came in voices and languages not alien exactly, but moved a step or two sideways, away from their daylight sound and meaning. So salt became the king of stones, and water mead of the desert; a dress was a veil, and sleep the bringer of sadness. The voices spoke of the future, or redrew the past. They spoke to unseen personages – now, you know you must not torment Yetemegnu in this way. I have told you and told you not to, and you must listen. They promised comfort, or delivered warnings.

Other times the woman behind the curtain would reply in her own voice. Ayzosh Yetemegnu, take heart. These things happen, and they pass. Or, are you sure you have not offended your zar? Have you prepared your offerings in exactly the manner I told you? A black chicken, remember, throat cut and left outside –

Yes, madam. Of course. I will do that tomorrow. And she would back away and take her place on the floor among the women, who welcomed her in. Looking about her, listening as others confessed their hopes and their troubles to the voice behind the curtain, breathing the close air, she began to ease.

Eve had thirty children, everyone here believed, and when God came calling she feared for them and hid fifteen of the most lovely away. But God knew she had hidden them, and decreed that henceforth they would remain in darkness. Zars, beautiful creatures with no toes and holes in the centres of their palms, were their descendants, who haunted the woods and glades of the northern plateaus. Like humans, the fallible children of light, they lived in families, they ate and drank, they quarrelled and they loved, but they used tree stumps as tables, gazelles and antelope were their cattle, elephants and lions their steeds. And when they wished to come into the light they chose a human as their horse, and rode him or her – though most often her – into submission.

Some could be malevolent, but most could be persuaded into kindness – and, importantly, protection – with gifts of perfume, with jewellery, with sacrifice. Over the years she had discovered that she had more than one zar: Gumay Lelé, a zar from Tigré; Ateté, an Oromo female; a Muslim zar who caused her to speak in something approaching Arabic, and to puff, awkward, at cigarettes; Shanqit, female servant of Rahelo, greatest of the female zars; and above all, Seyfu Chengeré, commander of the right guard of the zars. His gift was to understand medicine, his totem was a lemon tree, and his ally and adviser the Christian saint Gebrè-Menfes Qiddus, who travelled the world with sixty lions and sixty leopards and once obtained forgiveness for the people of Arabia by hanging from a cliff for thirty years, allowing his body to be pecked by birds. Gebrè-Menfes Qiddus lived on Mount Zuqwala and secured God's mercy for Ethiopia by placing on one side of a scale straw and weeds, for her sins, and on the other honey, milk and wheat, for her penances. Each zar required offerings – particular shades of sheep or hens or goats, intuited by reading coffee dregs, or the fat marbling through meat, or arrangements of thrown-down pebbles. Zars were always hungry, and especially hungry simply to be remembered – hence the daily coffee ceremony. Sometimes there was so much meat in the house and so many strictures about who could and could not eat it that her daughter, discomfited, felt she could not invite school friends round.

Yetemegnu understood how challenging all this was for her children, but she also knew this space, and these rituals, were necessary to her, that they were generous, and supportive. Calming. And so she kept going to the zar-house down the road. There food was served, roast grain, or bread, or small morsels of spiced tripe and liver, blessed by the zar-doctor. Sometimes a bottle of araqi was passed around. Coffee was poured and drunk with reverence, for in this room each tray, with its round-bottomed coffee pot, its delicate china finjals, and its incense burner, raised, occasionally, or, if it contained certain aromatic roots, passed under the women's skirts, was dedicated to an individual zar. Sometimes a tincture of healing leaves was handed round too.

The drum slipped into a regular rhythm, slow and deep, an exaggerated heartbeat. The sounds from behind the curtain began to change, the grunts and pants becoming cries, faster and faster, the drum climbing with them. A scream. And silence. Finally the curtain was drawn aside to reveal a very old woman, so tiny and thin it seemed impossible such great sounds could have come from her.

After a pause the drum began again. One of the women rose, and began to dance. Another joined her, and another. Some led with their shoulders, others with their hips. Some jumped and skipped, others bit and snarled. Some danced vertical dances of atonement, others dances of mercy and acquittal. Occasionally they beat themselves, as a rider might a too-slow horse. Sometimes women chewed on crushed glass, others walked through embers. Her half-sister, from whom she had received her first zar, in that dark hut long ago, used to claim that when possessed she could put a glowing coal into her mouth and cause no injury. All of them, echoing the behaviour expected of brides on their wedding night, attempting to hold in balance a struggle for freedom and inevitable capitulation, of joy and pain; all ending by moving faster and faster, until they collapsed, exhausted, and those who were watching them clapped and clapped.

The room was murky with smoke. She watched until she felt her conscious self melt away and her body take over, felt her limbs begin to move and her head begin to thrust, and then she knew nothing, nothing at all, until she heard the clapping, and slowly, slowly, opened her eyes to look about her. Her eyes met other eyes, encouraging, eyes holding her close, holding her safe.

Some of the lamps had guttered out. A couple of women bowed and left. Her body ached, but her mind was still.

MEGABIT
THE SEVENTH MONTH

Dry. Burning of bush to keep locusts away, and to clear new fields. Peaches and other cultivated fruits ripen, and garden crops are harvested. Caravans hurry to complete tours before the rains. Cattle pastured far into the bush.

When the spot appeared in the sky, she watched it as though it was she herself who was bringing it in to land, noting every tip of its wings, feeling every drop in altitude as though it was occurring somewhere in her stomach. She watched impatient as men scattered the sheep, sticks flashing about their narrow muddy rumps. Watched as the aeroplane bumped to a stop on the grass, the propellers slowed from blur to blades, and the stairs were folded out.

Ilililil!

The airport was full of people, relatives, schoolmates, friends, all come to welcome her son.

There he was! Dark and thin and looking for her.

Ilililil! Ilililil!

She bowed to the ground in thanks, and then she was holding his head tight to hers, burying her face in his neck, breathing him in, kissing him as though she would swallow him whole from love and missing him.

He smelled different. Of sharp air and a sweet clean fruit she could not name. Of new materials – plastic and cloth, cold metal and milled wood.

Ilililil!

They climbed into a taxi. A line of cars followed, and then more people followed the cars, walking and singing along the long road from the airport, past the army barracks and the hospital, the careful rows of Italian-built farmhouses, over the bridges and the rocky riverbeds, answering onlookers who asked who is this, why the dancing? 'It's Aleqa Tsega's son, become a doctor, he's back for a holiday, from five years in Canada.'

That first day she fed everyone expensive wheaten bread, because, she said, she had been instructed to do so in a dream. She moved among her guests, enjoining them to eat, delighting in the atmosphere, the festival of it, but always, whoever she was talking to, whatever questions she answered, she kept an eye on her son.

He held himself so carefully. He bowed and greeted people and kissed their cheeks and answered their questions, no one could fault his manners, and yet it was as though something within him was sitting away from the hubbub, staring. And he was not eating, she could see he was not eating. Eventually she took his best friend from school aside. Here. Take this food, and take him somewhere else, somewhere quiet. Make sure he eats. So his friend drew him away, into the street and up to the castles, where it was still among the trees, where bees buzzed in and out of their hives, and ivy pushed blind fingers through tumbled-down walls.

The priests danced, she always claimed, for eight days. They danced until their turbans came unwound and drooped and bobbed along with the drums. Minstrels came, and made every-one clap and laugh. Her cousin bought a sheep, and she bought a sheep, many guests brought something, there was food and beer for everyone.

But he had to leave, eventually, had to go back to Canada to finish his training, and after she had seen him off, she returned to the routines of her widowhood.

She had had her favourite homiliaries copied onto parchment and after communion each morning a deacon came to the house to read them to her, just as her aunt had been read to by the man who would become her husband, all those years ago. Sometimes, placing a stool at a decent distance from him, she would sit to listen; sometimes she went about her work, stirring a popping pan of shirro, chopping onions and garlic fine, while his voice ran the familiar words together, curving and lifting like the lengths of cotton on the weaver's loom down in the market. When after an hour or so he closed the wooden end-boards and slipped the book back into the battered case that smelled so strongly of incense and hide and decades of handling, she felt a pang of loss. But there was cooking to be done, rent to be received, a store of firewood to be built up for the rainy season, and so she would feed the deacon and send him on his way.

Alemitu had moved out of the main house, to two rooms in the compound that she had made her own. And she had met a man she liked, a trader from Debrè Tabor who bought low and in bulk – teff, niger seed – and waited for the price to go up before selling it on. She found him good company. He was an accomplished cook, an ally whom, importantly, she herself had chosen, and when her mother murmured about unorthodox arrangements she snapped, 'It's none of your business! I am grown, and I am not a nun. I must live my own life.'

Yetemegnu's youngest, Maré, was still at home, too, and when she was not in school or doing homework (about which Yetemegnu would tolerate no dissent) they worked together in the garden or around the house, polishing, spinning, cooking. She hoisted Maré, far too big for this now, onto her back and went out to plant cucumbers, because, she said, to have a child strapped to your back at the time of planting would make the fruits strong and healthy, the absurdity of it making her laugh so hard her daughter felt the shaking deep in her own body. As the rainy season approached, women began to bring in firewood, eucalyptus from the land at Bisnit. The piles rose up the walls of the house until she called Maré and handed some to her, saying, here, take this to our neighbours, so they may have a taste too. She did the same with berberé and with shirro – take a share to all our neighbours, especially those poorer than we are, and then we will be ready for winter.

And as they worked they talked. Advice, of course – respect your elders, no gossiping ever, give generously, in secret if necessary, don't sit with your legs apart like that, stand up straight, work hard – but confidences too. How much she still missed her own mother. I loved her so much and she died so early. She called me nigisté, have I told you that? My queen. Yes, Nannyé. And before she had her fill of me, before I had my fill of her, she died. Yes, Nannyé. Her daughter listened, and began to understand the gilding need gives to memory, and also that what was being offered was everything her mother had never had, and had not yet had the chance to give.

So when Maré got love letters at school – she was spirited and pretty and scandals and suspicions notwithstanding came from a good family – she trusted her mother enough to bring them back and read them out. And when she fell in love herself she was, unlike her friends, allowed to bring the boy to the house and talk to him there.

There were strict instructions, however. Be careful, Maré, my honey. He is a nice boy and serious and I want you to be happy, your sisters and brothers have been successful in their education and I want this for you, too. She said nothing about Alemitu, although she was beginning to regret, deeply, her earlier refusal of training in Addis. Be careful, Maré. Please.

Then when the sun went down and after they had eaten they would put the radio on, and Maré would get to her feet and dance – dances from Gondar, from Tigré or from Shewa, watching her mother all the while, for the moment when she could take her hand and persuade her, laughing and demurring, to join in. Minstrels often stopped by, knowing they would be welcome, and Alemitu, and for a couple of hours they would be united in movement and music and lamplight.

One day Yetemegnu called to Alemitu and handed her an umbrella. Not the kind of umbrella with a canvas shade that, rare in her childhood, had become ordinary, the kind of thing any lady might carry to ward off sun or rain, but a ceremonial umbrella, shivering with tassels and braided with gold. Take this to Abunè Aregay – a church outside Gondar, possessed of a tabot revered above most other tabots, and a reputation for encouraging fertility. Alemitu did as she requested, and nine months later gave birth to a baby girl. Tigist, or Patience.

MIYAZIA
THE EIGHTH MONTH

*Autumn. Light rains. As the soil softens ploughing begins,
especially after Lent. Peasants trade for new iron plough
points. Prices of meat, grain, pepper, double in
preparation for the end of the fasting. Second big
wedding season after Easter.*

Outside the mules were waiting, but she refused to be rushed,
checking and rechecking she had everything she needed. Good
clothes, of course, she had to underline her position in the world,
but not too good. A small cloth bag, hung around her neck in easy
smelling distance, containing garlic, crushed chalk and rue.
Lentils in the saddlebag. Also dried globe thistle, a piece of iron,
and a needle. She closed her mouth tight, then pulled her shemma
up to her nose, into one nostril of which she had stuffed a small
wad of rue, for extra protection. The rest of her face she set stern;
she could not risk betraying the slightest glimmer of joy, or of
pride – anything that might excite envy. At last she was ready, and
they were clattering up the street.

Beneath her the mule's back moved, narrow and warm. It was
not an especially healthy animal – it seemed to her it could not see

very well – which she had pointed out to the drovers when they arrived, but not vociferously enough to overcome their insistence that it was fine, a king among beasts, everything was fine. There were no reins, either, so she gripped the pommel as they left the Shinta and began to climb. The mule huffed and shook. She leaned forward to help it, as its forelegs bent nearly double and its hind legs stretched out, dislodging stones that rang against each other or rolled down the hill behind them, making jagged music as they went. Ché! shouted the drovers, slapping its rump. Ché!

The shepherds' voices carried across the slopes. A calm night? A calm night. It was a long time since she had been to Gonderoch Mariam, and it still held potent memories for her, of flight and of pain – but also of many days after the war, when the whole family had made their way up the slopes to inspect their land, checking on the administration of the church (in the case of her husband) and visiting friends while the children played, kicking footballs across a flat summit that seemed to extend into the sky.

They tethered the mules outside the church, and she went in to kiss its walls before making her way to a low constellation of huts nearby. Chickens scratched. A couple of children looked up at her. Where are your fathers? A wide-eyed pause, and a scurrying exit.

Their fathers, the farmers who tended this land, had sent their produce down to her house in the city for as long as she could remember. When, on recent trips to Addis, she had noticed the banners the students carried on their now-yearly demonstrations – Choice to the People! Land to the Tiller! – it had not really occurred to her that these attitudes, so foreign to everything she knew and assumed, might have traction in the provinces. But the tithes from Gonderoch Mariam had dwindled, and now some of the farmers were refusing to send anything at all, which was why she was here. Why were they doing this, overturning the natural order of things?

She knew some of the farmers, needing to supplement their income, were blacksmiths too, and their wives were often potters. Hence her elaborate precautions before setting off, her globe thistle and her rue, her garlic, her iron and her needle. For craftsmen in general, but Jewish blacksmiths especially, were believed to possess the evil eye, and the malaise caused by the evil eye was strong, stronger than anything a doctor or even a powerful zar could cure. Blacksmiths, who were said to inherit this sorcery from their fathers, and to depend on it for their ironwork skills, were thought to become hyenas at night, or to ride hyenas across the countryside, looking for victims. In the day they hid in plain sight, as men and women with shining, searching eyes; most did not in themselves mean any harm, but could not curb their power, which searched for those with outstanding qualities: wealth, beauty, gladness. In times of war or epidemic Jews suffered terrible retributions from neighbours who blamed them for their trouble. Already separated by religion and economic opportunity, they were further shunned and called Falasha, which means exile.

She feared them, but she needed them too, and so when the children's fathers came she was diplomatic and cordial as well as on her guard. 'I have sent enough of my harvest,' one said. 'There will be not a kernel more.' His beautiful young son, questioned by her, looked expressionless back at her and said not a word. 'He's deaf,' said another. 'Deaf. He doesn't hear you.'

When she returned a few weeks later she was told this was not true, that the boy could hear perfectly well. She was there for his funeral, because there had been an accident. The boy's father, who forged ploughshares, had been working his bellows when a pipe exploded, sending white-hot shards tearing through his son. After the ceremony, after the fragments of bone and flesh had been retrieved and interred, she spoke to the father, expressing sorrow.

'God has quarrelled with me,' the man said. Yes, she said, God does many things.

They went back a different way than they had come, shorter but interrupted by a wide ditch running with water. The other route was perfectly serviceable, she said to the drovers, but her mind was elsewhere and again she was overruled. It's fine, it's a good mule, they said, it can leap over.

But it could not. Its hooves slipped and she watched as its receding back crashed into the water, throwing her clear. I'm not wet, she thought. What luck. And then the pain arrived.

When they had first wrapped the arm up no one had thought to straighten it. Blood had collected at her elbow and every time she moved her arm she screamed. Eventually they called the local healer, who made a poultice of ground fenugreek he changed every three days. The bones healed slowly, and they healed askew.

But however tempted she might have been to stay at home, to cradle her weakened arm and run gentle fingers over its new contours, the courts had their own rhythm, and very soon she had to slip back into her layers of dresses and walk not to the courthouse, but to Bisnit, the land just up from Ba'ata that her family had ploughed for decades, and that the monks had claimed.

She stood in the cool groves and watched as the men she had brought with her cast about for landmarks. The sun, hot and strong out in the open, here only dappled the ground. They could hear the cool trickle of the stream as it worked its way between the rocks, and birds, singing among themselves. The court recorder stood slightly apart, writing in a notebook, face giving nothing away.

Alemantè was there too, though he was now no longer a boy. When the court in Addis Ababa had ordered her to find as many witnesses as she could, people who might remember where the boundaries of her land lay, he had come back to Gondar to help her. The two of them had waited in the marketplace while these strange men had assembled about them. Alemantè, having just finished a law degree – a choice prompted by the endless childhood trips to the courts with Yetemegnu – had wanted to know exactly what qualified them and had quizzed each in turn. I came here as a child, with my father, came the invariable answer. He was conducting business and wanted me to learn how it worked, so I followed him everywhere. He gave me roasted wheat or chickpeas as a reward. Now the children were men, and wandered among the eucalyptus trees, remembering.

Here, they said. Here from this creek that goes through the electricity generator, then runs into the Qeha, to there.

Or, See this spur of land? Here.

When the court recorder had finished writing they all dispersed, and a few days later she and Alemantè took a lumbering bus to Addis Ababa.

She stayed with him this time, in a small flat he had rented – as did, at various points, and for various holidays, three of his brothers, two of her daughters (Tiruworq between nursing school and compulsory countryside service, Zenna before starting as a teacher), Teklé, a serving girl and her baby. Being at home most days, waiting for court appointments, she took charge of feeding whoever was there; making sure that even if there wasn't quite enough at least Alemantè, who had just won a job at the recently constituted ministry of land reform, was always well fed.

They piled into the small rooms, kind, polite, loving – eat, upon my death eat, no, not that, this; don't go out there! The sun will make you ill, you'll catch your death of cold! They agreed their mother needed protecting, and held to their policy of not telling her anything about which she might worry. But they hid things from each other too. Molla, for instance, had at one point lived with Alemantè for months, but apart from frequent late nights and, one tense evening, the appearance of a poster of Lenin on their wall, gave little clue to what he was thinking. Years of watching his mother had made him instinctively anti-monarchist even before he became caught up in student politics and the demonstrations that were occurring, with increasing confidence, each year: for civil liberties, educational reform, or the people of Vietnam; against concentration camps or the government of Rhodesia, and always, always (even if the students had never been to the countryside and did not know a single farmer) for Land to the Tiller. He took a teaching job and during the day preached equality along with geography and biology; at night he learned Russian, and read a translation of *Das Kapital*. And then he too won a scholarship. He travelled to Gondar to say goodbye to his mother and then, via the Sudan (to camouflage his final destination from the authorities) to Moscow, to study medicine.

It wasn't too long, this time, before she received her court summons. There were three judges sitting that day, and in front of each of them arguments written by her nephew. Not for him the grandstanding, the physical flamboyance, the circumlocutions and resonant phrase-making of the traditional lawyers: his arguments were self-consciously, proudly modern – numbered, clear, and brief.

And not for the judges either, as it turned out. She had braced herself for months of appeals and counter-appeals, but they never came. In under three months – an unheard-of speed – the judges called them back to announce their decision. They had found nothing to disagree with. The land was hers.

GINBOT
THE NINTH MONTH

*Light rains, spots of fresh green grass. Storks fly
north. Women prepare fuel for the rainy season:
deadwood, and sundried cow dung coated with
mud. Caravans hurry home from Sudan.
Fishing in rivers. Children sing of the country's
wellbeing to storks, men and women picnic outside,
celebrating the birthday of Mary.*

She dreamt the emperor came to her on a small black mule, glossy
and groomed. Back straight, eyes calm, hands quiet over the
animal's neck. His cape fell in heavy folds onto a European saddle.

Follow me, he said. Take your mule and follow me. So she
picked up her mule's reins and followed him.

They travelled until they came to a high building. The emperor
dismounted and, unsaddling his mule, pointed it westwards. It
ambled toward the horizon and dropped out of sight.

Follow me, he said again. And so she followed him, up into the
tower. The walls were of marble, and they dazzled her. The stairs
were without end. At last they came to a door. Low, oddly placed,
it reminded her of the entrance to the undercroft at Ba'ata. The

emperor opened it and walked in, and then it was as though the ground sank beneath his feet, engulfing him.

She looked about her at the smooth floors and featureless walls and realised she was alone. Panic overtook her. What if someone came? They would think she was a thief. They would arrest her. They would kill her. She began to run, up steps, down steps, up, down, till her breath roared in her ears and the floor billowed at her, mocking.

Eventually a window appeared and she rushed toward it and looked out. It was high above the ground, but craning she could see underneath it a pile of manure, such as was sometimes left outside her own walls at the end of market day. Carefully, waiting after each move until her trembling had subsided a little, she climbed out and, praying, slid down.

She did not know how long she had been sitting on the roadside when a man appeared, walking toward her. In one hand he held a carafe and in the other a pitcher. Please! she called out. Please, I can't move, I am shaking so much I can't move.

He crossed the road. 'Healer of the world, what have we here?' And he put the vessels down and began to help her along the road, supporting her when her knees gave out, holding her when she could not move at all.

They came to a low house. A woman sat outside it baking bread. A pale woman, stately, unsurprised to see her. 'Take care of her,' the man said. 'I entrust her to you.' And he went on his way.

The woman smiled and held out a piece of fresh bread. She accepted, cupping both hands and bowing, but she was shaking still, looking about her for the pursuers she was sure would appear, and she could not eat.

*

Her son's best friend came to see her, and to read her a letter. I have found my life partner, wrote Edemariam. Frances is a fellow student, a doctor like me. She is Canadian. I know you will have imagined me marrying an Ethiopian girl, but it is a long time since I lived in Ethiopia. I like her, and I hope you can give us your blessing.

He would later hear that at first she had been upset by this, that yes, a mixed marriage had not been quite what she had in mind for her eldest son. And she had worried about the girl, how would she manage, in a country so different to her own? What if it didn't work, and she went home, taking any children with her? She would never see her grandchildren again. But when finally she dictated her reply it read, 'I will love whom you love, and I give you my blessing.' She knew there were few Ethiopian girls who had an education to match his, although the empress had set up a school for girls, who also now went abroad to study. Her own husband had had various ideas about who Edemariam should marry, and it would usually have fallen to her to implement them, or even to supply her own suggestions, but who was she to do that, in this new, modern world?

She travelled to Addis Ababa and the airport to meet them. She kissed the ground in thanks for their safe coming, she kissed her son and she kissed his new wife. She sat in the foyer of a hotel and smiled at her daughter-in-law, and her daughter-in-law smiled back. A pale girl, pale even for a foreigner, with long dark hair piled in a braid on top of her head and thoughtful eyes. She reached over and patted Frances on the thigh, comfortingly, realising as she did so that her inability to speak English was making her behave as though the girl was deaf. So she said, Enaté, ihité. Ayzosh. My mother, my sister. Take heart. 'Egzieryistiling,' said Frances. Thank you. Then they looked at each other and smiled again, unable, for the moment, to proceed.

Back in Gondar she haunted the house of a neighbour who had acquired a telephone. When her children called – often, as requested – she would hurry across the road, go down on her knees, and give thanks for the bare fact of the squat machine, praising its inventor and the God who had inspired him. Everyone laughed at her for it – Nannyé, we have had telephones since Menelik's day! – but she didn't care. It made her so happy. She felt the same way about cars and about aeroplanes – all things that brought her children close to her – suppressing the fact that they also took them away again. Ritual dispatched, she would take the handset. 'Allo? 'Allo?

And so she tracked their doings, Tiruworq nursing, Zenna teaching. Edemariam and Frances working, almost immediately, at the Hailè Selassie I Hospital opposite Genetè Le'ul, which had, after the attempted coup, been given with much fanfare to the university. She worried when Edemariam was sent to Harar for countryside service, worried about his wife walking through the streets, about the attention that followed foreigners, the shouts of 'Ferenj! Ferenj!', worried about her living alone, and enjoined Teklé to look after her until Edemariam returned to establish a teaching unit at the free hospital, St Paul's. She approved when Edemariam went to see his benefactor Theophilos, now elevated, after the death of Basilios, to patriarch of the Orthodox Church, and laughed, aghast, when she heard her new daughter-in-law had shooed a meddling visitor from a hospital room, only to find the visitor was Princess Tenagnewerq, and not much minded to do as she was told. When Edemariam and Frances were invited to one of the many receptions celebrating the wedding of the crown prince's daughter she wanted to know every detail – how on the third or fourth evening of this vastly lavish affair they sat at the bottom end of a long table, sipping fortified mead and waiting for a banquet that took so long to arrive at least one foreign diplomat had to be carried out insensible. Or of how, in celebration of the emperor's eightieth birthday (though a strong rumour circulated that he was in fact about eighty-five) the entire city had been smartened up, with banners, coloured lights, and bright-painted hoardings to hide the slums. How every national institution had been required to hold its own party, and how for weeks his subjects journeyed in from every corner of the country, bringing gifts.

But she also had an instinct for the things they did not tell her. Her children grew used to fielding expensive long-distance calls. I have had a dream. This son – or that daughter, or the other – is not well, is hurting, or is unhappy. Something is wrong. No, Nannyé, nothing's wrong, don't worry. Yes. There is something wrong. Tell me!

Sometimes she was placated, other times she stood her ground. Once she was staying with Zenna in Debrè Zeit, a bustling, jacaranda-lined town an hour's car journey south of Addis, when she dreamed of a woman so old and so tiny she looked like a child. She even spoke like a child, or perhaps an adult's sense of a child, a high-pitched, indistinct lisp. The little zar looked at her, then raised her arm and rang a bell. Follow me. So she followed until they came to a big house surrounded by a verandah. Through a doorway she could see Edemariam's wife lying on her stomach on a long bed. She was facing east and reading a book. Edemariam leaned against the verandah railings. His face was dark, ashy, and down his left side, from head to toe, hovered clouds of grey and white, like cotton wool with the seeds picked out. The little zar pitter-pattered up to him. Then she grasped the upper edges of the cloud and pulled it down until it lay like an unhooked curtain, moving in folds about his feet.

Zenna? she called. Zenna? I had a dream about Edemariam. He was very ill, and now he is better.

What a good dream, said her daughter.

When she next saw her son he admitted that yes, he *had* been ill, hospitalised with tuberculosis contracted on the wards at St Paul's.

The zar healed you.

No, Nannyé. You really shouldn't believe in them. It's backward, embarrassing. I had injections. I had medication.

More births, then.

After five years of not always patient waiting she persuaded her eldest to shoulder another ornate umbrella and return to the monastery. Nine months later another baby duly appeared. Nine ililta: a girl, Elsabet, on Christmas Day.

Then one rainy season she boarded a bus to Addis Ababa. On the roof and about her feet and on her lap pots and packages, baskets and cases. For weeks she and the household had been preparing everything necessary for a woman after a birth: thick porridge swimming in butter and chilli, cracked wheat, honey and oatmeal gruel.

My mother smiled and thanked her, and refused all her largesse. But you must! Upon my death you must. How will you regain your strength? My mother stayed firm. There were further shocks. She put me to sleep on my belly rather than on my back. But she will stop breathing! She will die! It had not been an easy birth, might, in fact, have resulted in at least one death if it had not occurred in hospital, but within days my mother was standing and walking about. Sit! Sit! Go back to bed! You must heal, you must take it more slowly!

Within two weeks my mother had left the house to visit a friend. Everyone was appalled by that. Forty days! Forty days are yours, in which to rest, to be cared for and waited on, and for forty days you must remain at home. And yet here she was – my mother took me outdoors, into the sunlight. But in the sun, wailed my grandmother, a baby is open to pestilence, to relapsing fever, pneumonia, the evil eye. Oh it's not good, it's not good, Mary, mother of God protect her, it's not good. Ayzosh, said my mother, who was learning Amharic, and especially important words like this; ayzosh. She will be fine. And she kept on taking me out of doors.

When I was eighty days old they all woke before dawn and carried me round the corner to Abunè Theophilos's church, St Mary's, where in a basement room I was stripped of my white dress and plunged howling into a font of cold water. My grandmother chose my baptismal name, Iqibtè-Mariam.

During the rains, when the heavy clouds lift for a moment and from underneath them sunlight reaches long fingers across mountain slopes and down into the valleys, it catches on trees: sycamore fig, juniper, wild olive and rosewood, fern pine, cedar and eucalyptus. And the fresh, rounded leaves of young eucalyptus catch the sun and reflect it back, each leaf a jubilant, quivering mirror. As they age these leaves lengthen into blue-green points that slide and dance, drawing traceries across the sky. If the women have not been through for a while, gathering fuel for cooking fires, the ground is littered with them: fresh leaves, tough but malleable, that bend and bruise and sharpen the air with their aroma. In the dry season they brown and harden, then craze and shatter, scattering brown confetti across red earth. And everywhere lies the fruit – rough conical hats with ridged seams, scored underneath with four-point, five-point, six-point stars.

The first eucalyptus trees to take root in Ethiopia were planted around St George's Cathedral during the reign of Menelik II. They grew quickly, their trunks slim and strong, and people saw, first with surprise and then with an acute sense of utility, that after they were cut down for firewood – fast-burning, fantastically hot – new trunks rose from the roots. Within a few years eucalyptuses were everywhere, their muted silhouettes radically softening and saving a capital that had long ago burned through its native trees and faced abandonment.

By the 1970s the country felt as though it was built on them: eucalyptus supported the houses and shaded the churchyards; it held up the electric cables and spread like webs across rapidly rising buildings. Long eucalyptus branches, wrapped into heavy bundles, were strapped to women's backs and bent them double in the dawn, men fashioned walking canes and used them to drive oxen or to sling across their shoulders as they swaggered down the roads. Eucalyptus smoke flavoured the food, the beer, the very air; a copse of mature eucalyptus meant a child's books, a coat for school, a dowry or funeral dues; it was insurance, and it was wealth.

When she was awarded the land at Bisnit she had been awarded all the trees as well – those that stood on her land, and those she had planted on the land that now belonged to the monks. Now she cut them down and sold them, and hosted a wedding party for a relative's daughter she had taken in to raise, in the same way she had taken Alemantè. For years she boasted about what a brilliant wedding it was. We had a huge bull slaughtered, and a cow, the sauces came out right, the beer was so good, hundreds of people were fed.

SENÉ

THE TENTH MONTH

*Heavy rain, fog, storms. Fishing season. Children taught
to swim before rivers turn dangerous. Intensive ploughing,
seeding by broadcasting of barley, teff, wheat. Garden
crops seeded and fenced.*

Later she would say that was when the loneliness came – when the
wedding was over and her youngest had boarded a bus for Addis
and school. It had been there always, waiting, glimpsed out of the
corner of her eye as she stirred split peas or picked through
cracked wheat, watching from the ceiling as she lay in bed, her
daughter breathing beside her, whispering through the trees when
she went out into the garden to tend her pumpkins. Now the loneliness flowed into the open, dimming the sky, an ache that kept
her awake into the small hours, refusing to be kept at bay.

This was to some extent a matter of perception, however, for it
was also true that Alemitu was still at home, and now there were
Alemitu's two daughters. There were house servants and neighbours and the usual trickle of relations from the countryside,
dropping by to sip beer, be fed, exchange, after long silent inter-
vals, titbits of family news, and then go on their way. There was

church, and there was the zar-house, and then there was, after so
many years, more building work. When her son returned from
Canada he had decided she could not keep living in a house with
earthen floors and no modern plumbing; in gratefulness for all
she had done for him he would, before he bought a home for
himself, build one for her. They had discussed it off and on for
months, but when I was born, and my mother's parents arrived
from Ontario to help her, my Canadian grandfather, a skilled
carpenter, drew plans for it and suddenly the house her husband
built was being knocked down around her, the irregular stones,
held together with mud and grass and, some insisted, raw egg,
were being prised apart and carted away and replaced with breeze-
block and cement, cast-iron piping, a toilet, a sink, a bath.

Along with the new house came, for the first time, her own
telephone. She never tired of the sound of it ringing, of picking up
or being handed the handset, which reminded her of a loom, and
speaking into it. 'Allo? 'Allo? How are you? All well? Everyone
well? The children? The house? Work? All well? Now tell me. Tell
me all your news. She had the radio moved into pride of place in
the new living room. Ish! she admonished everyone, three times
a day. Ish! And sat down to listen.

Two days after Christmas the price of petrol went up by 50 per
cent. Not many ordinary people had cars or trucks, but that was
not the point: those who did tended to be nobility, government
employees – and merchants, who hauled grain and spices, bolts of
cloth and teetering piles of kitchen implements into the cities and
now promptly put up their prices. The Saturday market filled with
disgruntled traders and disgruntled shoppers and all the talk was
of prices, of prices and of unrest. In Addis the taxi drivers went on
strike, making of the main avenues loud static rivers of white and
blue.

She took to the phone. Are you safe? Are you well?

I'm well, Nannyé, we're all well, everything's well. Ayzosh.

The government raised salaries and lowered the price of petrol, but still the cities chafed. Buses were stoned in the capital by running groups of boys, and by taxi drivers outraged the bus drivers had not joined in their protest. She could see those buses, smell them almost – listing, puffing, overfilled and underserviced, the lion leaping across their sides (the emperor owned a stake in the company) both an easy target and a mockery of their actual rate of progress.

Are you safe? Are you well?

I'm well, Nannyé, we're all well, everything's well. Ayzosh.

There was trouble in Eritrea. Gondar being a garrison town, the soldiers from Azezo drank in the bars and lounged in the Piassa, and darker news began to filter through, of mutinies at outposts in the south, of a famine in the north that had been ignored.

The prime minister and his cabinet resigned, but it seemed to make no difference: the strikes spread through Addis, then through the major towns and cities. There was a two-day general strike, which paralysed everything. The new prime minister called out the Fourth Division. Initially they did as they were told. Then they joined in. Postal workers and transport workers went on strike. Schools were closed. The prostitutes demonstrated, and lay priests. Thousands of women marched, and tens of thousands of Muslims. What do *they* want? Equality. Laughter, at the very idea of it, but also the feeling that something was shifting, splitting open.

Are you safe? Are you well?

I'm well, Nannyé, we're all well, everything's well. Ayzosh.

During Holy Week her daughter-in-law's parents came to visit. My grandmother had been concerned for days about how she would cope, but her daughter-in-law's mother was gentle and kind, a former nurse, and her father was a United Church minister, interested in ecclesiastical architecture and the long Easter services, which made her feel more at home with these people with whom she did not share a language, and with whom she had to communicate in bows and smiles and gestures to eat more, eat, eat, please eat, why will you not eat?

Ten days after they left the army arrested the former cabinet. In a broadcast the following morning the army claimed they had done so on orders from the emperor. They threatened retribution for strikes and against 'trouble-makers in the civilian population'. But weeks of intermittent losses of power, water, telephone and radio contact made clear the warnings had not been entirely successful.

Are you safe? Are you well?

I'm well, Nannyé, we're all well, everything's well. Ayzosh.

The military took charge of all radio and TV stations. 'We, the armed forces, police and militia, would like to notify the public that we are ready to take the necessary action against the detained cabinet members, and at the same time would like to express our loyalty to the emperor and the Ethiopian people at large.' Five days later 'we' had become the Coordinating Committee of the Armed Forces, Police and Territorial Army. Two days after that they declared their guiding slogan: Etyopya tiqdem. Ethiopia First.

Nearly every day now it rained. Rain drummed on the corrugated-iron roofs, drowning out the radio, drowning out speech. It roared through the alleyways and tumbled down the streets in brown torrents. The market traders covered their wares with heavy sheets of plastic and huddled, sodden, in the corners of their stalls. Children sent on errands shivered from stone to treacherous shining stone. Hailstorms swept in over the mountains, ripping through the pumpkin leaves and depositing bulwarks of white against the walls. Lightning scrawled across the sky and thunder arrived so suddenly, so loudly, that she hid under her gabi. They laughed at her. Do you think you'll escape it like that?

Visitors were stranded, waiting for the onslaught to ease. Rumours brewed and festered.

They said that in Addis there were soldiers in the streets, tanks, roadblocks.

They said no one could understand what the emperor was doing. Why did he not act?

Every radio broadcast attacked the ruling classes, the rich, the landowners, and, by tacit extension – even though the Committee always made sure to declare its support for him – the emperor himself. And with each attack the question became more urgent. Why did he not act? What was going on?

Then her half-sister the nun came to see her. 'They've arrested him!' Who? 'Asratè Kassa!'

The rainy season wore on. Everything smelled of curdled woodsmoke and old sweat and damp cotton. They washed clothes, trying to catch the brief hours of sunshine, but nothing would dry. People wrapped themselves up against the cold in layer after layer of rapidly greying gabis and shemmas and shawls, drawing them tight around their shoulders and over their heads, burrowing down to wait for spring. The voices on the radio grew more strident, more bold. The people must rise, end tyranny. They must take their true place in their own country. The Coordinating Committee of the Armed Forces, Police and Territorial Army became such a familiar phrase, so much a part of everyday life, it became just 'the Committee'. The Derg. 'Who is this prince, Derg?' people asked, reaching for jokes to still rattled nerves. 'We never knew the emperor had a son by that name.' It was not that many didn't agree with at least parts of the criticism: everyone had complaints – about bribery, nepotism, tithes, taxes. But to do anything more than hope Hailè Selassie would finally, finally declare a successor? That was, for most, unthinkable.

The Derg abolished the Crown Council and the emperor's personal court. Memory rushed in on her then, of all those months in the press, in the sun, of the little man standing attentive at his desk. All those people who had waited, like her, for months and for years, waited for a word that might change their lives. All gone. A sudden feeling of vertigo, of unfamiliar air. A deep need to crouch down and curl in, like a woodlouse when the stone above it is removed and its home exposed.

Are you safe? Are you well?

I'm well, Nannyé, we're all well, everything's well. Ayzosh.

The radio was on all the time. News bulletins had always been dominated by the emperor's doings, and that was still the case. But factories toured, countries visited, modernisations promised were now replaced by accusations that the emperor had deceived his people, had abandoned them by fleeing in their time of need, had sold public holdings for his own profit, that above all when his people were starving, walking from Wello to Addis and dying in their thousands, he had feasted on chicken stew and fed his dogs choice beef. They called him king instead of king of kings, and the increasingly lurid, increasingly bold whispers were of corruption and opulence and human sacrifice. And when on New Year's Day she sat down to listen to Abunè Theophilos's annual address, his usual prayers for the Lord of Lords, the Conquering Lion of the Tribe of Judah, the Elect of God and all his issue were replaced by intercessions for the Derg.

Emperor Hailè Selassie I was deposed just after dawn the next morning. His parliament and his constitution were void, said the newsreader. The crown prince would be enthroned as soon as he returned from medical treatment abroad. Any protest would be dealt with in military courts, against whose judgements there could be no appeal. Elections would occur in due course. Etyopya tiqdem!

That night, and then night after night, she prayed, not always sure what exactly she was praying for. For the emperor, as for a fallible father who had done much good as well as ill? For Princess Tenagnewerq, arrested at her home on New Year's Day, as for a relative, a kind patron, or an impersonal royal? For her suddenly unrecognisable country? Was it hope she felt, or fear, or grief, or some impossible combination of all three? Fear for her children, certainly, for their health and for their safety. She prayed that Mary would always intercede for them, would remember that every one of them was named after her, wherever they might be.

With the rest of the country she listened and watched for signs and portents. The rains eased and rainbows flung themselves across the skies. At Masqal, which marks the beginning of spring as well as the arrival of the True Cross, the emperor was not there to light the bonfire. Long after dusk, when the flames had eaten through its heart, observers noticed that it collapsed not toward the palace, presaging a good year, or away from the palace, presaging a bad one, but in on itself, into a glowing, smoking, hissing pile of charred sticks and embers.

And when she finally slept, she dreamed, deep dreams from which she woke with her heart pounding erratic, or dreams entirely forgotten, then surfacing as false memories, full of foreboding. Dreams of cities and pageants, of the wild four corners of the world, dreams of her children.

Are you safe? Are you well?

I'm well, Nannyé, we're all well, everything's well. Ayzosh.

She dreamed of a country, this country, but somehow more beautiful than anything on earth could ever be. A drystone wall and next to it three trees, two eucalyptus and a juniper. It seemed to her that old Ras Kassa had come from abroad and stood there under them, preparing to cut the juniper down. And that she was begging, begging him in the name of the law, in the name of Hailè Selassie, in the name of the God above both of these that he should not. If you cut it down our boundaries will disappear. This is not your land, he replied. And she cried back, it *is* mine! It's mine! But he turned away from her, and rather than using an axe, took hold of the trunk with his bare hands and began to rock and pull at the tree. Dust and berries and sharp narrow leaves rained down on his head and settled on his shoulders but still he pulled and shook, niqniqniqniq. Slowly, the whole tree began to lean eastwards and then to fall, faster and faster, until with a whoosh and a crash it was horizontal, branches splayed like a skirt across the scratched ground. Directly he moved to the eucalyptus trees and did the same again, and the air was vivid with the astringent smell of torn leaves and her vision filled with the ragged splinters of broken tree trunks.

Then, one bright summer morning, some weeks after the emperor's deposition, there was another announcement. Yesterday, said the announcer, it became necessary to execute 'all those responsible for the misery of the masses through abuse of authority, maladministration and judicial malpractices'. Execute? But there had been no mention of a trial! Military court or no military court, surely justice had to be seen to be done? Name after name followed, ten, twenty, thirty, forty, fifty, sixty, three former prime ministers, ministers, vice-ministers, one of the emperor's grandsons, the commander of the imperial bodyguard, the chief of police, provincial governors, various members of parliament, the chief of staff of the territorial army, the Derg's own chairman, Le'ul-Ras Asratè Kassa –

She stopped listening.

*

They had made an effort. The dining table had disappeared under the bowls of stews, of vegetables, of cottage cheese and injera and rice and salads. From the sideboard rose a city of bottles: beer, mead, Fanta, Coke, golden towers of Johnnie Walker. The guests sat around the edges of the room in mismatched chairs commandeered from the bedroom, the servants' quarters, the neighbours. The rented house her son and his wife had moved into was comfortable but not large, and the living room felt very full.

She looked around, at her daughters bending to serve each guest in turn, at her sons, and tuned in and out of the polite hubbub. Conversation these days was full of new words. The word revolution, for instance – though it was almost always accompanied by an old word, to drive home the experience of it. The revolution exploded, people said to each other. How else to describe what had happened to them? There were more baffling words, like socialism, whose meaning seemed to shift according to who was speaking. Simple words like 'change', which seemed suddenly to carry a weight vaster than they could support. Words yoked together in alien and apparently indissoluble ways, development through cooperation, Land to the Tiller – a dictum that had been acted upon only a few months ago: all rural land, said the Derg, now belonged to the people. Henceforth, and for the first time in the country's history, there would be no tenancy, no tithes – and no private landowners, thus erasing at one swoop an entire ruling class (and dashing the hopes of any farmers who dreamed they might one day till their own fields). And so teff and wheat and peas would no longer arrive from Yetemegnu's acres at Gonderoch Mariam. There would be no compensation. After the initial shock of the news she decided she would not miss it that much. The amounts had been dropping for years, administration was a challenge for a woman alone, and anyway, grain was cheap and easy enough to buy in the market on her doorstep with the money she earned in rent.

And always, everywhere, at the end of every broadcast, every official communiqué, repeated and repeated until promise began to assume the lineaments of threat: Etyopya tiqdem. Ethiopia first.

She looked around the busy room, and was in turn aware of being watched. She knew this lunch was in her honour, and also that they had something to tell her – she could feel it, in the glances of expectation and covert care. Was it a death? No one should learn of a death alone – but you would not host a party to announce a death. Prison? Had someone been imprisoned? So many had been. She looked across at the wife of the former pensions minister. A Betè Israel from Simano, her father's village in Dembiya, he had been arrested along with the other ministers, but so far he had not been shot. Her moment of satisfaction at being avenged had shaded quickly into horror, horror and sorrow for all of the executed sixty, everyone had felt that; horror and sorrow and a waiting dread of what might happen next, if the Derg were capable of this.

What did they want to tell her? Her bowels turned then and she rose and threaded her way through knees and children and occasional tables and out into a bedroom, where she drew the door to and prostrated herself, praying to Teklè-Haimanot, to Gabriel, to Mikael, to Mary above all, that the news should not be too heavy to bear.

When the main course was cleared away, Edemariam rose and seemed about to speak. Then, obviously nervous, he sat down. A few minutes later he stood again, looked at her, made to speak, sat down. She could see he was in agonies and willed him strength. The third time he remained standing.

'Mother, my mother. You're strong, aren't you? You're brave?'

Of course I'm strong, she answered, comforting him, but fighting, also, her own fear and her weariness of fear. Didn't I prove that, through all those years in court, fighting for my land, my houses?

He winced. And then he said, Nannyé.

Yes, child?

The houses have been taken.

What houses?

The houses in Gondar. The rental properties.

The fear poured out of her, leaving her empty and shaking.

What? That broken-down stuff in Gondar?

All taken, he said, his eyes hunting hers. He told her later he had invited guests to soften the blow; too many people had suffered fatal heart attacks on hearing they had lost their livelihoods.

She held his gaze.

Let them. As long as they don't take my children.

And again she dreamed. She dreamed she was in her big house in Gondar, that she had visitors. That the chat was proceeding as it usually did, How are you? Well, and you? Well, and your children? And your husband? All well, thanks be to God. A convivial silence, and frankincense, drifting.

Then someone said, 'Did you hear?' Hear what? 'The emperor is gone. He's been called away.'

Her heart dropped. But my children! My children! I gave them to him for safekeeping!

Nannyé. She woke to a voice in her ear. Nannyé. Ayzosh. What is it?

Sweat trickled down her forehead and ran between her breasts. Sweat swallowed her up and spat her out, stranding her, shivering, among the bedclothes.

What does it mean? What does it mean? What is to happen?

When, days later, military music buzzed through her radio and a voice told her Hailè Selassie I was dead, it was as though she already knew.

HAMLÉ
THE ELEVENTH MONTH

Heavy monsoon rains almost daily, sometimes tornadoes.
Cold. No visiting or outdoor work due to treacherous
waterways and washed-out paths.

Teff tripled in price and chilli peppers followed suit. There was no wheat to be found, anywhere at all. The servant girl would disappear for hours, looking for a twist of sugar to put in the morning coffee. When Yetemegnu returned from Addis she had gone to the leader of the new local urban association and pointed out that now they had taken both her land and her rooms her only income was a tiny stipend, a percentage of the rent now collected by the government. If that was what everyone got, she could manage, she supposed, but what of her eldest, who could not hear? What was she supposed to do?

Eventually they gave Alemitu a small plot of land on which to build her own house. What did we do to deserve this? she asked Alemantè when he visited. Why was our property taken? We didn't steal it. We are not the nobility. He had never seen her so bitter. He was by then permanent secretary at the ministry of land reform and administration, but he had nothing to say to her, because while he had agreed with the nationalisation of rural land he didn't believe in the most recent measures himself.

Blame and counter-blame, mistrust and division, flourished. The merchants were hoarding, said the provisional government; they were profiteering. They had to be punished. There is nothing coming from the farms, the merchants countered. The government in its zeal has given state farms over to the workers, and the result is confusion and dropping yields. There is little to sell. In Addis the Derg executed a few merchants as examples, imprisoned more. Students who only a few months ago had marched into the countryside singing of revolution, hoping to bring it to every farmer and cowherd, saw they had been tricked; that the aim had been to get them and their ideals out of the way while the provisional military government consolidated power that had nothing to do with the people. So the students traded their khaki uniforms and soft caps for mufti and slipped back into the towns. Seditionary leaflets passed from hand to hand. Ras Mengesha Seyoum of Tigré, married to the emperor's granddaughter, had led a guerrilla army into the northern hills as soon as the sixty had been executed. He accused the Derg of betraying the people, and argued for the right to form a true democracy. Argument, conducted in increasingly alien language – feudalism, imperialism, capitalism, proletariat – bloomed in the newspapers, on the radio, in the streets; it grew between siblings, between parents and their children, between childhood friends, and separated them. In Addis my father watched as Alemantè and Molla, newly back from Moscow, traded acronyms like blows and joked that they might as well be talking about pharmaceuticals, for all the sense he could make of them.

They arrested Abunè Theophilos. Never mind that he had led one of the few official bodies to respond wholeheartedly and constructively to the famine, was popular and known to be progressive – he had ordained bishops without the Derg's permission. The patriarch had buried grain in the basements of churches, they claimed. He amassed wealth and kept women. How risibly transparent, she thought. He is holy! she said to anyone who would listen. He is like a saint! He helped me so much and was so kind. She imagined him in prison, felt it viscerally, as a violation of everything she held to be immutable, foundational, right.

Alemantè disappeared. One night he was staying at her son's house in Addis; the next morning he was gone. Eventually they heard he was in Tigré, then, many weeks after that, in the Sudan.

Even she who could not read could see that the newspapers – which everyone scanned, urgent with the need to divine the contours of this new world – looked different. They were full of lists. What are they? she asked. Lists of the arrested, lists of the dead. Lists of enemies of the revolution, now eliminated. These so-called enemies conducted their own eliminations in response, and although they did not publish lists, many saw, or heard, or smelled the results: bodies tumbled into badly dug graves or snagged in the debris under bridges. Bodies lying on roadsides, in the alleyways behind back fences. Women milling around entrances to prisons, self-effacing but intent on not having to take the food they carried home again, because that would mean the prisoner was dead.

Then one morning a face in those papers. Sloping-back brow. Large nose. Dark pockmarked skin. Small eyes. Obviously short. Like a slave, she thought, thrown back onto the divisions of her childhood. But entirely unslave-like he stood on a stage before microphones, mouth wide, throat corded with shouting. In one hand a piece of paper, in the other, raised high above his head as though launching a spear, a glass bottle full of dark.

Major Mengistu Hailemariam had that week led a coup against other rivals for leadership of the Derg and after a shootout emerged triumphant. 'As a result of the determined and decisive step taken Thursday by the Provisional Military Administrative Council, our Revolution,' he bellowed into the crowd, drawing them into collective responsibility, 'has, in keeping with the demands of the broad masses, advanced from the defensive to the offensive position. Henceforth we will tackle enemies that come face to face with us … we will arm the allies and comrades of the broad masses without giving respite to reactionaries and avenge the blood of our comrades double and triple-fold …' He smashed the bottle to the ground. It had contained water, dyed red.

It was evening and they were sitting in candlelight – as so often these days, there was no electricity – when Alemitu's partner came to see her. The flame guttered as he opened the door and shadows quivered across the walls. Her heart rapped at her ribs and she was sharp with him. You scared me! What are you doing, creeping about in the dark?

'I just wanted to show you,' he said. 'Look what they gave me.' He held a rifle, and a pistol.

She knew why he had been given these things. A son from a previous union had joined an anti-government cell, then thought better of it and run away; in fact Yetemegnu had helped him run away, telling him to leave the city, even putting him on a bus. The other cell-members had come for his father then, and threatened to kill him unless he gave up his son. Alemitu's partner refused, and went to the authorities to ask for protection, which they said they could not provide. But they had handed him these guns. Don't show them to a single person! she had said to him. Wrap the big gun up and hide it under your bed. Make sure no one sees the pistol, but hold it close. Then if they come to kill you, you can take at least one of them down with you. She looked at his fearful face. Don't show anyone, do you hear me?

When, within the month, the local association defence squads came searching for arms, banging on doors to seize weapons which, in many cases, people had always owned, weapons of Italian vintage, even some, older, once hidden from Italians conducting similar searches; weapons kept as naturally as one would a seasoned walking stick or a fattening sheep, for self-defence or just as evidence of being someone who counted, she handed them her own little Colt. A few days later she went into the municipal offices and, shaking, asked for it back. I am a woman all alone, she said to the soldier behind the desk – everything, it seemed, being run by soldiers now. I want to protect myself against thieves. But he was unmoved.

Arms seizures became seizures of anything 'counter-revolutionary': printing materials, cameras, building materials, petrol, typewriters, food stores – and of anyone who, in a much-repeated mordant précis of Mengistu's wild speech, 'dared to think, or encouraged others to think'. Or, in fact, of anyone with whom the defence squads personally disagreed. The lists in the papers, of the supposedly counter-revolutionary dead, were no longer of names, just numbers. The revolutionary fire flamed bright and hot, pitiless and undiscerning. On May Day thousands of students marched in protest against the public's disenfranchisement. They were massacred. Those who fled were overtaken and consumed, shot as they ran, shot where they hid, shot in their homes and their safe houses. Parents were required to pay for bullets lodged in their children's bodies before they could take them away to bury them. The unclaimed and the yet to be identified lay in the streets where they fell and the hyenas came down from the hills and feasted on them. May the Red Terror spread far and wide, said Mengistu, and the Red Terror obliged.

The Red Terror came to visit her, running and stumbling down the hill into the stalls of the Saturday market in the form of a teenager taught by anti-government cells to shoot and to kill. He had missed a target, grazing only an ear, so now he was running, scattering stones, slipping on straw and donkey dung, chased by a gathering mob shouting 'Get him! Get him!' And then they did get him, someone taking a stick and aiming a great blow at the teenager's waist, another at his head, splitting it open. And then the anti-government cells came looking for his killers, and soldiers came looking for the anti-government cells, and in the mêlée those who had been watching Alemitu's partner saw their chance: a man he knew slightly, a relative, came up to him to give him a hug, and while he was thus distracted, the relative's accomplice shot him.

She was having her hair buttered and braided for Epiphany, incense was burning and the coffee boiling when they took her into the back room to sit her down and tell her, but she had hardly been able to absorb the news when the square in front of the houses was swarming with men and guns. 'Bring them out! Bring them out or there'll be trouble!' But everyone had shut their doors, their gates, their compounds and sat shaking and silent inside.

'Bring them out!' A crash, and two men were standing in her living room and a gun was pointing at her. One of the men swung his leg and kicked over the incense burner. His shoe came off and flew across the room. 'Stop!' said his companion. 'That's enough!' She found her tongue, and also, to her own surprise, a rush of reckless anger. Kill me if you want, but really, isn't it enough? Isn't it enough that my son-in-law is dead? A tense pause, and then the muzzle of the gun was pointing at the floor, and she was looking at their departing backs.

Worry, always in the wings, became a constant presence. Worry about her children in Addis, where the conflagration was fiercest.

Worry about her daughters, her granddaughters. She worried herself sick about Molla, now serving as a doctor on the northern shore of Lake Tana and one terrible day arrested as he worked. She went to see him, carrying food to the prison gates like all the other women all over the country. He had been in leg irons for days and was covered in lice and fleas. He tried to make her laugh, as he always tried to make her laugh, to reassure her he would not be there long, but all she could do was cry and pray and glance over and over at the windows, sure soldiers would shoot through them at any moment. In the event he was right – he was released quite quickly – but her children stuck even more zealously to minimum news after that. It would be years before she discovered much about his next posting, how as a battlefield doctor in the war against Somalia he spent part of his time training local women in midwifery and family planning, preaching against female circumcision in an area where women suffered total infibulations rather than the partial cutting of the highlands. How one day he had been sent, with a ninety-soldier guard and the general's gun, to inspect an area where cholera had taken hold. How on their way there they saw a giraffe blown to pieces by a roadside bomb and on their way back vultures dropping out of the sky. How as they approached they ran into an ambush, a bitter rain of bullets that killed twenty of the soldiers in his escort. How he spent the night applying tourniquets and comforting the wounded and praying that his two-year-old son would not have to accompany a coffin back to Addis. All this he kept to himself, protecting her.

But from some things she could not be protected. Hangings had begun again, from the branches of the sycamore fig, and floggings, convicted criminals bound hand and foot and thrown to the ground in the market to receive forty or eighty lashes, each lash counted out by the watching crowd. Political prisoners were treated differently. Nearly every day, it sometimes seemed, military vehicles full of young men and women arrived at the prison round the corner. They were not executed in broad daylight but after nightfall, after the curfew, and they were shot rather than hanged or whipped, machine guns chugging through their awful cycle, the sound impossible to block out however she might cover her head and stop up her ears, every scream and echoing edge loud and louder because the air was so still and everything so silent, listening. Then she would throw herself across the ground in distress. Or she would tie her girdle tighter about herself until she could scarcely breathe, trying to stop the pain. Sometimes she would return to her senses knowing time had passed but not where she had been.

The next morning she would keep her grandchildren home from school because she knew the bodies would be thrown out on the roadsides for parents to retrieve, and she did not want them to see. There was no one who did not know someone dead, disappeared, dying; no parent, it seemed to her, who did not spend their days remonstrating, please, don't join them – whichever side 'them' happened to be – don't fight, don't retaliate, think, please think. But their children replied that it was the dawn of a new age, and parents knew nothing. She could not rid her mind of stories she heard, stories that could have been just gossip, or just as easily true, of sons killing parents who tried to stop them, of a young man captured, his arms spread wide like Yesus Christos, his feet tied together, his hands pinned to the ground with nails, his eyes spilled out of his head and his entrails cast across the ground like a sheep's. Hating to be alive, hating creation and everything in it, she took to her bed.

And then someone transferred Abunè Theophilos from Menelik's palace, where he was under arrest, to Asratè Kassa's old home, and strangled him.

NEHASSÉ

THE TWELFTH MONTH

Heavy rains. Fishing in muddy, turbulent rivers.
Small amount of sowing. Cattle taken upland.

… WHEN I PRAY UNTO THEE, I THE SINNER AND TRANSGRESSOR,
DO THOU O VIRGIN MARY, INCLINE THINE EAR TO THE VOICE OF
MY PETITION.

The rain pounded on the roof and rattled through the guttering.
She could feel her voice moving up her throat, the movement
of her tongue, warm breath across dry lips, but the words themselves were lost against the splattering dance of raindrops on
concrete.

… AND HEAR IT AND NOT BE IMPATIENT WITH ME, BUT WITH A
SHINING HEART AND A PURE MIND ACCEPT THE WORD OF MY
MOUTH.

The floor was strewn with grass. Among the blades lay tiny white
flowers of sorghum popcorn, some still intact, others crushed by
feet on the way to the kitchen, or into the sodden yard. The thin,

handle-less cups were not yet washed out. Coffee grounds collected at their bases like rubble in a dry streambed.

IT IS NOT ROBES OF HONOUR MADE OF PURPLE, AND SILK, AND CLOTHS DECORATED AND ADORNED WITH DIVERS COLOURS ... BUT I LAY OUT MY SOUL BEFORE THEE IN THE PLACE OF GLORIOUS APPAREL DECORATED WITH GOLD ... AND TO THEE I ... DECLARE ... MY ... MY ... SIN ...

She stumbled. For an instant the words on the page before her became just marks again, discrete letters, as they had been for most of her life. The hide binding felt rough in her hands. A breath. A look up, and then down, and the words came back into focus, the letters joined together again to form meaning, their dark little legs and arms and heads gesturing toward worlds she could walk into on her own now, at her own speed and her own behest.

I HAVE FOUND THEE A REFUGE FROM THE CORRUPTION WHICH IS ON EARTH, AND FROM THE PUNISHMENT WHICH IS FOR EVER. I HAVE FOUND THEE A REFUGE FROM THE LIONS OF THE NORTH, WHICH ROAR MIGHTILY, AND SNATCH AWAY WITH VIOLENCE, AND HUNT THE YOUNG AND SHOW NO MERCY TO THE OLD, AND GAPE WITH THEIR MOUTHS TO SWALLOW UP THEIR PREY.

When the literacy cadres came knocking she had looked into their zealous faces and told them she could already read and write even though – apart from being able to sign her name as Edemariam had taught her, and which she duly demonstrated – this was not true. Caught between the demands of the revolution and the far older demands of tradition and respect for their elders, they bowed and left.

Then one morning, which she would for ever after remember was the festival of the Mount of Olives, she had gone to retrieve some cash from its hiding place between the leaves of one of the children's school primers. She had always thought the letters beautifully printed, appreciating them as familiar objects, and on a more abstract level as tools in her children's and grandchildren's education, but this time words leapt out at her. Sentences. The hyena ate the cow. Kebedè bought some basil. It was as though an angel had alighted on her shoulder and with a sweep of his wings unveiled a new dimension. Every day she looked again. The dog walked down the road. Kebedè, who seemed a busy sort of fellow, brought home a mortar and pestle. She felt radiant with amazement and exhilaration as the world lifted, expanded, spread itself out before her.

I HAVE FOUND THEE A REFUGE FROM THE WOLVES WHICH DO NOT SLEEP TILL THE DAWN, WHICH SEIZE AND CARRY OFF AND LEAVE NO SHEEP UNTOUCHED, AND SPARE NEITHER THE YOUNG GOAT NOR THE LAMB. I HAVE FOUND THEE A REFUGE FROM THE FACE OF THE BOW AND FROM THE MOUTHS OF SPEARS AND SWORDS, AND FROM EVERY INSTRUMENT OF WAR.

She often thought, after that, of Yared, who as a boy was not a good student. His teacher often beat him and so he ran away, into the countryside. He was resting under a fig tree, listening to the sounds of the birds and the waterfalls, when he noticed a little worm trying to climb the trunk. The worm would get partway up then fall back, try again, get very slightly further, fall back. For hours he watched it, until at last it gained the canopy and began to eat the figs. It was a pivotal point in his life. He returned to school and eventually became the first composer of Ethiopian church music, and a saint. So she thought of him and of the worm and after her own vision kept trying too, again and again and again. The simple sentences became longer, more complex. She read of Golgotha, and picked a stumbling path through the psalms of David.

I HAVE FOUND THEE A REFUGE FROM THE HANDS OF ALL MINE ENEMIES, AND FROM THE HANDS OF THOSE WHO HATE MY SOUL. WHO CAN STRIKE TERROR INTO HIM THAT PUTTETH HIS TRUST IN THY NAME?

She read all the chapters in the Book of Job and the Book of the Praise of Mary; she returned to a favourite story, that of the saint Christos Semra, who became a nun out of contrition after losing her temper with a servant and accidentally killing her. Who after her time at Debrè Libanos prayed for many years in a hole in the ground out of the sides of which spears protruded, piercing her each time she moved. After decades of supplication Christos Semra began to feel she ought to attempt to reconcile Satan and God, and so she travelled to the lips of Sheol and called lovingly: 'Satnayél! Satnayél!' The devil came, and began to drag her down into his abode by her eyelashes. The archangel Mikael had to split the abyss open with a bolt of lightning to get her out, but with her Christos Semra brought ten thousand souls, who leapt with gladness into the light. She had read the story so many times she felt she knew this female saint; she longed to go to Tana Qirqos, the island on which Christos Semra, having reached the grand old age of 375, was buried, and which, on the twenty-fourth day of the twelfth month, became a place of pilgrimage. She had flown over it so many times on her way to Addis. If only she could set foot upon it.

I BESEECH THEE, O VIRGIN MARY, THAT THY PRAYER MAY BE TO
ME A SHIELD OF HELP, AND THAT THE POWER OF THE ARMS OF
THY FIRSTBORN MAY COME DOWN TO DELIVER ME. LET THE
POWER THAT BREAKETH THE MOUNTAINS COME TO
OVERTHROW MINE ENEMY. LET THE POWER THAT ROLLETH UP
THE HEAVENS COME TO CAST DOWN HIM WHO WOULD OPPRESS
ME.

All those who had testified against her husband began to leave the
world.

Witness one died when surgical stitches in his abdomen, an
Italian job that had lasted for decades, finally ruptured. His intestines unravelled and he expired in agony.

Witness two died in prison having stolen an ox.

Witness three died when his body erupted in pustules and his
flesh began to peel away.

Witness four had a stroke.

Witness five died blind.

And she felt that she stood taller.

THOU WAST NAMED 'BELOVED WOMAN', THOU BLESSED AMONG
WOMEN. THOU ART THE SECOND CHAMBER, IN THAT THOU
WAST CALLED 'HOLIEST OF HOLIES', AND IN IT WAS THE TABLE
OF THE COVENANT AND ON IT WERE TEN WORDS WHICH WERE
WRITTEN BY THE FINGERS OF GOD.

In the middle of the twelfth month the festival of the
Transfiguration, a festival of flames and torchlight, of boys singing, going from door to door minting verses, begging for alms.
Boys who each year grew younger and younger as their brothers
and friends disappeared. The luckier boys were hidden by their
parents or smuggled out of the country. The less lucky – the poor,

mostly – were pressed into the army by a government desperate to stem losses in the wars against Eritrea and Somalia. Not that this was the official story. Officially there were only wild war-boasts and vast slogan-littered march-pasts; few knew much for sure, but the garrison at Azezo was huge now and nothing could stop the rumours of desertion and confusion and terrible morale, of battles where hundreds of men were sent charging over minefields, then hundreds more sent after them, forced to march over their comrades' shattered bodies. There were soldiers everywhere, soldiers often so spooked, so trigger-happy, that I remember rounds of machine-gun bullets fired into the heart of a tornado, where they joined the tree branches and sheets of corrugated iron flying like dark leaves of paper through the tarnished sky. They had to land somewhere, of course; my grandmother overruled my curiosity, shoved me and my cousin into her wardrobe, then, with the house help and a neighbour or two, stood in front of it, forming a human shield while we crouched among soft white dresses that smelled of incense and woodsmoke and limes and tried to imagine what was going on. They were aiming, the soldiers said when the storm had passed, to kill the devil in the wind.

She like everyone else saw the grey-green trucks that crawled through the streets, boys standing in the open backs, gripping the wooden slats or slumped down against them, faces hidden away. Like everyone else she knew the trucks stopped to let soldiers leap out and grab more boys off the street, boys who attempted to run away, boys who thought they were too young and thus immune, but they were not immune, because few checked or cared whether they were tall for their age or even what their age might be.

And she mourned them as though they were already dead, because so many did not come back, and if they did it was so often as paraplegics or with missing limbs, lost either in war or because they had shot off toes, trigger fingers, when they could think of no other method of escape and they did not want to or could not join the multitudes in prisoner-of-war camps or those who walked for days into the Sudan, stumbling into heat and endless exile. So many boys mutilated in this way the sight attained a kind of twisted normality.

AND BECAUSE OF THIS WE ALL MAGNIFY THEE, O OUR LADY, THOU EVER PURE GOD-BEARER. WE BESEECH THEE AND LIFT OUR EYES TO THEE, SO THAT WE MAY FIND MERCY AND COMPASSION WITH THE LOVER OF MEN.

There had always been a measure of watching; the emperor had had his spies, his efficient and secretive reporting systems, but this was of a different order. Mengistu's little eyes followed her across shops, post offices, banks, looked down at her from billboards as she walked the streets. Often he was joined by a trinity of bearded white men – MarxEngelsandLenin, said the children, running them together into one entity, but she was none the wiser. Soldiers manned checkpoints and patted down anyone entering or leaving official buildings; militiamen in grubby cotton gabis carrying Kalashnikovs lounged at urban association gates or strolled down back streets, looking, listening, threatening. Everyone knew and if they did not know suspected that behind the police officers, the militiamen, the soldiers in uniform, there were secret security forces, that death squads still roamed the shadows and the prisons overflowed, even though the Terror was theoretically over. And then there were the private watchers, the friends who proved to be less than friends, the siblings who found themselves on opposite sides, the strangers who were slightly too interested, the wide spaces that opened up alongside phone conversations because of the invisible, listening presences. Everyone knew not to say anything of import if they could help it. Or to look around and about and then around and about again before a rushed whispering – or thinking better of it altogether.

Because although there were words everywhere – shouting from newspapers, strung up across streets, blaring from loudspeakers mounted on jeeps, towering two storeys high on government buildings, spilling in six-hour speeches from Mengistu's mouth or from the radio and the television, where every single broadcast began 'Mengistu Hailemariam, Chairman of the Provisional Military Administrative Council, Chairman of the Council of Ministers, Commander in Chief of the Armed Forces, General Secretary of the Party, and (eventually) President of the Democratic Republic of Ethiopia' – they were not generally helpful words; they were not real news. 'Etyopya tiqdem!' (again and again and again). 'Revolutionary Motherland or death!' 'Marxism-Leninism is our guide!' 'Down with American imperialism!' 'Long Live Proletarian Internationalism!' 'The oppressed masses shall be victorious!' They did not explain – or perhaps explained all too clearly – why there was not enough food in the markets and the shops, why gunshots split the night, or why not everyone lying at the side of the road was sleeping.

THOU ART THE TABOT WHICH WAS COVERED ON ALL ITS SIDES WITH GOLD, AND WAS MADE OF THE WOOD THAT NEVER PERISHETH, AND THAT FORESHADOWED FOR US THE WORD OF GOD, WHO BECAME MAN WITHOUT SEPARATION AND CHANGE, THE PURE AND UNDEFILED DEITY, THE EQUAL OF THE FATHER.

The skies were blue, and stayed blue long into the time when they should have given way to rain. Carrion-birds spiralled down from the mountains, vultures, fork-tailed hawks, searching the ground lazily, diving suddenly, secure in their dominance. Brown earth turned yellow then dried into dust. In the Saturday market, under the castle walls, on the steps of the Italian post office, the hungry joined the halt and the limbless. Every day children and servant-girls queued for hours for the bounty of a few small loaves at 10 cents each.

As the tenth anniversary of the revolution approached, the rivers of words became floods, but they did not carry with them any mention of food shortages, of failed rains and stillborn crops. Red stars rose above the streets, and along them murals depicting the proletariat marching bright-faced and triumphant through rich wheatfields, hammers and sickles aloft. (While many of the proletariat in fact lived behind those murals, in slums.) Ethiopia could feed itself five times over, boasted Mengistu. It would be a breadbasket for the region for generations to come. She heard him on the radio and like many others believed him. Red flags flapped from every lamppost, and the pictures of Mengistu, Marx and Engels and Lenin, the shouting banners, proliferated. 'Workers of the World Unite!' In Addis, in front of foreign dignitaries fighting to stay awake, Mengistu gave a five-and-a-half-hour speech extolling the glories of his own achievement.

THOU ART THE HOLY GOLDEN POT WHEREIN THE MANNA IS HIDDEN, THE BREAD WHICH CAME DOWN FROM HEAVEN, THE GIVER OF LIFE UNTO ALL THE WORLD.

And then the starving began to arrive. They came into the Saturday market and when, at dusk, the traders packed up, they moved onto the low stone platforms on which the stalls had stood and settled down to sleep. Red and green and yellow lightbulbs, strung up in celebration, cast dim light on inert piles of torn gabis, hollow cheeks. Proud, they tried to work. They sold every scrap of wood they found in their path; the women offered hairdressing, domestic labour. Like many others she invited a few into her compound, asked them to braid her hair, and the children's hair, to tend the fire, and fed them. All over the city people began to hear of camps and feeding stations, of flies crawling over corpses, of the inescapable smell. Of limited food supplies reaching only the very worst-off. Of cholera racking the living – only no one was allowed to call it cholera. The few who gave the disease its true name (my mother, for instance, who spent her days on under-resourced wards filled with the vomiting, dehydrated dying) were accused of political sabotage, as was anyone who said the famine was anything other than a story planted by the CIA. In the newspaper they published pictures of the old famine, under the emperor – the scourge from which Mengistu had supposedly saved his people. She and Alemitu stared at a picture of tiny twins, all bones and vast eyes, trying to suckle at the flat breasts of their dead mother, and could not stop weeping. One of the women who came to her house had three children and asked her to care for the smallest during the days, while the rest of the family collected wood, tearing trees out by the roots to get as much fuel to sell as possible. For two years the tiny, sweet-natured girl played among the pots and pans and followed her to church, until the family judged they had saved enough to go home to Wello, where both the girl and her mother died.

THOU ART THE GOLDEN CANDLESTICK AND DOST HOLD THE BRILLIANT LIGHT AT ALL TIMES, THE LIGHT WHICH IS THE LIGHT OF THE WORLD.

Odd things began to appear in the market and on the otherwise mostly empty shelves of the shops in the Piassa. Small metal tins decorated in bright red and blue and white that yielded, once a hole had been punched in the top, thick oversweet milk. Rough sacks of rice with foreign writing stamped across them. No one really knew how to cook the fat white grains, and once they were cooked the results were so sticky no one particularly wanted to eat them. Teff became so expensive and hard to get that even those with a little income, like herself, began to consider mixing it with foreign wheat. The government, unable to deny the famine any longer, announced that anyone earning over 50 birr a month – not a large sum – had to give up a month's salary for the relief effort. Ration books were issued and the preparation of food, already a chore that occupied most of all days, took up even more time, as a member of the household was dispatched to wait in a long list- less queue for the month's ration of sugar, tea, coffee, matches. People grumbled and groused, eyeing the guards with their sticks and Kalashnikovs, exhausted in the sun, but most also knew this method meant everyone, regardless of their station, would get the same minimum amount to eat.

Daily life was navigated through thickets of edicts. Only local produce could be used, local soap, local salt (quite grey), local sugar. All government employees must henceforth wear suits of local cloth, sewn by government-assigned tailors and available in only three colours: a livid, shiny blue, light blue, or khaki (the fact that the dye itself was imported was conveniently ignored). Wednesday afternoons were for political re-education. No one could leave their homes during the curfew, from 11 p.m. to 5 a.m. Car owners were allowed just enough petrol to drive to work and back. On Sundays there could be no driving at all, except for half the taxis (odd licence numbers one week, even the next).

The world, already small, shrank further, turned inward.

[THOU ART] THE GARDEN OF DELIGHT, THE GARDEN OF JOY, WHICH IS PLANTED WITH THE TREES OF LEBANON, AND WAS PREPARED FOR THE SAINTS BEFORE THE WORLD WAS CREATED.

At last, after nearly ten months, the rains returned, sluicing down out of the heavens with the force of an answered prayer, drumming holes into the dry earth, each raindrop seeming to bounce back up, remaining whole for an unfeasible instant before breaking and releasing its bounty into the ground. And the earth drank it in, thirsty, and released a smell so fresh, so intense and intoxicating and welcome few could resist the urge to go out, to lift faces, chests, palms up to it, to dance, even, in gratefulness. The sound was everywhere too, racketing in the gutters, dripping from dulled trees, swirling around the stones in the path down to the market, seeking out holes in the roof.

Her garden bloomed. Every morning when she returned from church, before she went in for coffee, she walked out into it, welcoming the dog that bounded up to her, inspecting the exploring tendrils, the shy orange hides of her precious pumpkins. Then she would stand among the wide leaves and look about her. She and Teklé had planted as much as they could – potatoes, with their star-shaped yellow-centred flowers; tomatoes, all dusty red skins and sharp-smelling hairy stems; tender lettuce; tough, dark-green kale; coriander, sacred basil, maize, broad beans, rue.

THOU ART THE LADDER WHICH JACOB SAW REACHING FROM EARTH TO HEAVEN, WITH THE ANGELS OF GOD ASCENDING AND DESCENDING UPON IT.

They had planted flowers, white rian, black rian, red roses in which she would bury her nose, breathing in lungfuls of perfume. When they were at their height she picked and then dried them, hanging them upside down by a scarf tied about their waists. The roses lived, then, in her clothes chest, among the dresses and the shawls, releasing their subtle scent every time she lifted the lid.

She had planted henna and winter cherry; purgative vernonia and elders for curing sores; borages, which were good for colds, as of course were the eucalyptus leaves from the trees that now towered over all these things. They had been growing ever since she and her husband had planted them, over thirty years before, and now they were mature, their high grey-green trunks, their shivering leaves, fencing in and bordering the quiet with the sound of a river flowing through the sky.

THOU ART THE BUSH WHICH MOSES SAW BLAZING WITH FIRE, AND THE WOOD THEREOF WAS NOT CONSUMED.

She didn't keep many chickens – she maintained they ruined a back yard with their scratching – but she had a couple, naming them Pumpkin and Shashé, feeding them, splinting broken leg bones and watching, helpless with sympathy, as they struggled to lay eggs. Despite herself and her best intentions she always fell in love with her milk-cows, loved their long-lashed eyes, their patience, remembered previous milk-cows she had raised from calves, feeding them injera, burning incense for them, naming them too (Welansa, and then Enquay, Welansa's child), treating them as an extended part of the family and weeping for them when they died. She raised puppies, making nests for them of old dresses and spoon-feeding them milk, running distraught into the streets to avoid the exterminators who came round after outbreaks of rabies, having asked, first, for a grave to be dug to receive their handiwork.

But she did not weep for the sheep, who on high days and holidays joined her menagerie for a few days, baa-ing and butting, pulling against the ropes that held them, until prayers were said over their heads, their throats cut, and they were strung up by their hind legs to be gutted.

And so she would stand and look around her, and then her eyes and her heart would return to the pumpkins again, imagining the rich orange flesh, the sweetness of the stew she would make from it, thinking this was fine, this was beautiful, that whatever was raging outside her gates, here at least was a private Eden. And then she would turn and go indoors.

THROUGH EVE WAS THE DOOR OF THE GARDEN SHUT FAST, AND BECAUSE OF MARY THE VIRGIN IT HATH BEEN OPENED TO US AGAIN.

Her youngest daughters had long gone, to Addis first, and Debrè Zeit, and then they had scattered, one to Canada like her eldest son, another to Bulgaria, yet another to Czechoslovakia, where they learned languages she had never heard of and in them studied nursing, economics, agriculture. She fed on their infrequent letters and even more infrequent visits home, and turned her energies on her granddaughters. She had started quite early to take the girls overnight, arguing that Alemitu, asleep, would not hear them cry; now they often chose her company over their mother's, heedless as children are of the raw hurt they thus caused. The eldest ate every meal with her and settled in her living room to study – or to war about studying, and about seeing friends in the evenings. Education came first, there could be no dissent. Get away from here! she would command, when her granddaughter stayed out late. Get out! I'll feed you to the hyenas, you impossible child! But then she would relent, and open the door, and draw her granddaughter into a hug, and put out her favourite food. Come, come to me, it's all right. Just don't do it again, do you hear?

AND A VOICE CAME TO THE HOLY WOMAN, SAYING, 'O MARY.'
AND SHE ANSWERED AND SAID, 'HERE AM I.' AND AGAIN, THE
VOICE SAID, 'REJOICE AND BE GLAD, AND LET THY SOUL BE
GLAD BECAUSE THOU HAST FOUND GREAT GRACE WITH ME ...'
AND STRAIGHTWAY THE MOTHER OF OUR LORD SAW A GREAT
LIGHT WHICH MAN CANNOT DESCRIBE, AND THEN THAT VOICE
SAID UNTO HER, 'HENCEFORTH THY BODY SHALL REST IN THE
GARDEN AND THY HOLY SOUL IN THE KINGDOM OF THE
HEAVENS.'

In the evenings, in these the darkest months of the year, in the
days preceding the feast of the Assumption, she drew her family
even closer about her. After supper – no meat or butter or eggs,
because the feast was preceded by fifteen days of fasting – they
would wash their hands, their wrists, their feet, and then they
would sing, meditative, joyous songs in praise of Mary, songs she
had known since she herself was a small child. Songs so familiar,
so comforting, that they unspooled like memory.

AND THEN OUR LORD ANSWERED AND SAID UNTO PETER, 'SPEAK
UNTO ALL THE BEINGS OF HEAVEN AND COMMAND THEM TO
SING HYMNS AND PLAY INSTRUMENTS OF MUSIC WITH JOY AND
GLADNESS'; AND AT THAT MOMENT THE SOUL OF THE HOLY
WOMAN MARY WENT FORTH, AND HE TOOK IT AWAY INTO THE
TREASURIES OF THE FATHER. THEN JOHN STRETCHED FORTH HIS
HAND AND STRAIGHTENED HER, AND CLOSED HER EYES; AND
PETER AND PAUL STRAIGHTENED HER HANDS, AND FEET.

And then sometimes, when the feast of the Assumption was over,
the songs were replaced by others, from the radio, or from a flat
tape-recorder, or from a minstrel who might happen by, looking
for work. They would listen, and clap, watching each other, smil-

ing – who would move first? The children, generally, snapping shoulders forward, then back, forward, back, catching her eye, moving faster, and faster. And then she might take a small shot of araqi, or of tej, and wipe her lips, and stand, and they would smile wider, or laugh, and move back to make a space for her. And she would step into the circle, put her hands on her hips, then, keeping the rest of her body still, still, address her chin forward, so. And a sweep to the side, so. And a sweep to the other side, so. And forward again. Body controlled, eyes serious, while inside her heart filled and filled. And then she might begin to move her shoulders too, and they all shook their shoulders, faster and faster, hair flapping, breath gasping, necklaces rapping on breastbones, faces opening into laughter so wide they ceased to be aware of it, only that they were dancing.

AND THERE APPEARED A GREAT WONDER IN HEAVEN; A WOMAN CLOTHED WITH THE SUN, AND THE MOON UNDER HER FEET, AND UPON HER HEAD A CROWN OF TWELVE STARS.

The next morning she would go to church again, to stand in her favourite spot by the south door, and to bow until her forehead touched the ground and the dust of the floor caught in the back of her throat and the chanting of the priests filled her head with light and silence.

BLESSED AM I BECAUSE OF MY BELIEVING AND NOT BECAUSE OF MY RIGHTEOUSNESS AND PURITY. BLESSED AM I BECAUSE I PUT MY CONFIDENCE IN GOD AND PRAY THE PRAYER OF HIS MOTHER.

*

A few weeks after the royal women were released from prison she and her eldest son, his wife and a family friend bought a bottle of Black Label and went to visit them.

The shaded gardens were quiet and the big house still. Gone, long gone, were the queues of petitioners, the bustling retainers, the gravitational pull of unquestioned power. She looked about her and shivered.

They had been asked to come, but when Princess Tenagnewerq spoke it seemed to her she had not been recognised.

But Your Highness, I am Yetemegnu, Aleqa Tsega's wife! I am your daughter. Don't you remember me?

'Oh,' said the princess. 'Yes. But you are older and darker now.'

Yes, Your Highness, we are all much older.

Fourteen years for the royal women; the emperor's grandsons, imprisoned as children, not yet released. (The crown prince, abroad when his father was deposed, wisely declined to return.)

The next thing she knew Tenagnewerq was bowing down to her, low, almost to her knees.

Oh don't, please don't! Unsure whether to touch her, how to lift her back up. What is it, my mother? What is it?

'I heard you foretold my father's death. That you said I longed for a father as if I did not have one, and that you said it before he died. I heard that when you pray God answers your prayers. Pray for me, please. Pray that I may see my brother again.'

Of course, Your Highness. Please rise. Of course, of course, of course.

PAGUMÉ
THE THIRTEENTH MONTH

The week after my daughter turned two, I took her to see her great-grandmother. It was the middle of the rainy season, when every day brought towering black clouds and thunder rolling around the horizon. Rahel had practised how to say hello in Amharic, how to say ayzosh. She told everyone we passed in the sodden streets who she was going to see. Nannyé had made a huge effort, and sat on her sofa in a full white dress, fresh headscarf, and perfectly draped shawl. I had last seen her three years before, and had spoken to her on the phone many times since. Of course I knew she was failing – who would not be, at nudging ninety-eight? That was partly why we had come – but the spirit had always been there, the presence and the engagement. This was different. I saw at once the cruelty of extreme old age, how it strips away the basic faculties, of movement, of sight (her eyes began to fail barely a decade after she learned to read), but, worst of all, how it strips away people: the fifth of her nine children had died the previous summer, and it was as though that, finally, was the last thing she could bear. Her whole body spoke of exhaustion, of darkness and sadness.

And Rahel – stupidly I had forgotten that the only photograph of Nannyé she knew had been taken in the 1930s, under the Italians. I had not tried to explain how the young person now existed in an old one. She took one look, retreated to the far corner of the room, curled up into a ball on the floor and refused to come out. Ney, called my grandmother, gently, neyilign. Come, come to me. Rahel curled in tighter. Want to go home! Eventually she ran outside, into the garden. Nannyé sat silent as I chased out and in, divided between trying to keep the small one safe and wanting to comfort the older one, say hello properly, re-establish my own connection. What time is it now? Nannyé asked. What time is it? What time?

Over the next few days Rahel practically lived in the garden, and I watched her discover for herself the smells, the sights and textures, that were the warp and weft of my own childhood, watched her revel in them as though on some level she recognised them too. So she picked and sniffed her way through clumps of rue, of rosemary and mint and lemongrass, and balanced, as if on a tightrope, on the breezeblocks that demarcated the vegetable garden. She swung on the bars of the water tower and inspected the fuchsia bush that grew at its base, comparing tight-closed buds with the pale skirts and dark purple underskirts of full-blown blossoms. Even the column of glossy army ants that one morning inspected her in turn, assaying her limbs and scalp, biting her and sending her screaming for me, did not deter her. She said assiduous hellos to the smaller, kinder sugar ants and stared at a lizard as it basked blinking on the high wall before dropping, a quick sinuous wholeness, out of sight. She shucked corn, watched the yellow kernels blacken over coals that glowed and smoked and settled, then attacked them with her teeth. I rolled up her sleeves, tied a teacloth about her neck, and passed her slices of mango so lush and ripe the juice ran down her forearms and her chin shone orange. She stared as coffee beans, stirred on a pan over the fire, turned from green to red-brown to a lustrous black and jumped in surprise when they began to pop and launch their outer husks into the air, where they drifted delicate, see-through, like a beetle's wings. When the housemaid took advantage of a break in the clouds to wash Nannyé's clothes she stood close, riveted by the movement of the girl's wrists above the rim of the shallow metal tray, the slop of the water as she pressed the heels of her hands into the wet muslin, churning the soap suds into thicker, richer froth, slop, slop, slop, and sometimes an extra slop, right over the edge and onto the ground next to Rahel's delighted feet.

She began to come inside more often, bearing gifts: three coffee beans, a brace of bruised fuchsia buds, a muddy carrot just dragged out of the ground. At mealtimes she and Nannyé sat at opposite sides of the room, being fed. Nannyé had no appetite, every sip and mouthful was coaxed into her. Rahel, generally a hearty eater, was wary of strange tastes and grains and looking for any kind of deflection. Oh when the saints come marching in, she sang one day, conversationally, across the coffee table. Oh when the saints come marching in. Then, Happy birthday! recently learned and sung with gusto to her grandfather on the phone, to the pictures of Mary and Jesus on the walls, to the father confessor in his black and yellow robes, making his fortnightly visit. Happy birthday dear Nannyer, happy birthday to you. Raheliyé! said Nannyé, who had resisted proper conversation for weeks, replying to her daughters, to me, in groans only, or in lists of aches and pains. Raheliyé, I know your father, you know. He came to Ethiopia and I danced with him. And she had. After our engagement lunch she had stood and walked to the centre of the room, where a knot of us had gathered. And she had straightened her shoulders and put her hands on her hips, had raised her head, and then her chin was moving, sharp, controlled, like a bird's, forward, left, right, forward, and the drums were loud, and the fiddle and the lyre, and standing opposite my grandmother, hands on my own hips, responding with my best approximation of the (far easier) shoulder-shaking version of the Gondaré dance, I realised I was smiling so hard I thought my face would split. After a couple of songs Nannyé was persuaded to sit, she was ninety-five or so after all. But I could tell she didn't want to.

The day we left there was a hailstorm. Rahel watched the white globes, big as grapes, shrink to points in her hands and cried More! More! until she was brought entire cupfuls of translucent spheres. She ran in to Nannyé to show them off. Their two heads leaned in, absorbed, one in a cup, the other in the small face in front of her. The rain of the thirteenth month is holy, it's Ruphael's rain, said Nannyé, to the top of Rahel's head. When I was a child we'd tear off our clothes and dance through it singing. And every leap year, when Pagumé was six days long instead of five, all the children and the youths would rise before dawn and go down to the Qeha to swim. She picked up Rahel's hands, and kissed them. Oh they're so cold! Geté! My jewel! My heart, they're so cold! And to everyone else, What time is it now? What time?

Six months later I flew back to Addis, and then, with other members of her immediate family, to Gondar, descending through cloud and landing on tarmac edged with tall grasses, an acacia, and an old control tower, rusting. When I first came, twenty-five years ago, this was a flat field grazed by cows and sheep which had to be buzzed away before a landing. Now it was a busy modern airport, all dark asphalt, TV screens, gift shops. The watchfully indolent soldiers were a constant, however. Only the uniforms changed. Nannyé had known the regime was in trouble, had seen and wept for the young men who, fleeing a lost war, had washed up in Gondar market, beautiful young men, she said, princes with cracked feet and troubled eyes, angular with hunger. Many young people refused to eat in solidarity, she remembered, sending their meals out to the soldiers instead; others went from house to house, collecting food. She herself took eight in, asking few questions but feeding and feeding them until there was not enough left that day for her own family. But she did not initially connect all this with the voice that came to her in her dreams, ordering her to tell her rosary, to sing the kyrie and weep for three months. What's coming? she asked, distressed. What's coming that I have to do this?

The day before the three months ended a grandchild sent to borrow a mortar and pestle ran back empty-handed, gabbling with news. The guards have gone! They've unlocked the prison and gone! Immediately my grandmother began to gather clothes, food, intending to make for the best sanctuary she could think of, Gimja-bet Mariam. Where should I hide? What should I do? she asked a friend who had dropped by. Nothing, he said. They're already here. There had been days of fighting at Azezo, but none in the city. All through the following days the neighbourhoods heaved with stories of their entrance, led, some said, by a guerrilla riding a donkey; that some brought black cats and others black dogs, instruments of witchcraft; of wild-haired militiamen taking up stations at each corner of the prison, except that one of them – people made excuses to walk by, just to stare – was a woman.

We step out of the airport into glittering sunshine and a wall of wailing. Everything in me except my feet steps back, and I am weeping before I am aware of it. All of us are, terrible tears, at the same time that part of me notices how manufactured some of it is, particularly in men who, not generally permitted such things, look for the right sound and largely fail. Then we climb into a four-wheel drive and set off to meet my grandmother.

She is waiting at the crossroads. New-tarred road and a vast sycamore fig and drawn up on the verge, back pointed to Addis Ababa and front toward Gondar, a pickup with a coffin in the back.

What time is it? What time? It's late, people are waiting, we must go, we must go – And the cortège sets off, crawling toward her home. She's been away for twenty years, ever since she returned to Addis because of another jailing, of her son Molla on political grounds, but most things have not changed. The long low hangars of the army camps are still there, and Azezo feels as it always did, a bustling way-station to the airport or Lake Tana, to Bahr Dar and Addis Ababa, or west to Metemma and the Sudan border; a long raucous strip of a town, all focused on the main road. Though some of it might have surprised her – the rows of stalls filled with bright Chinese plastic, the lean-to teahouses, clientèle staring as we pass. Dozens of cars cross the Shinta. Dozens of cars cross the Qeha. Past the hospital, past the university, past men piling unfeasible numbers of eucalyptus logs onto garris, their broken horses waiting, scarred heads down, unmoving. Past the Italian post office, down the Piassa. Past the castle walls, jacarandas in full bloom threaded through the turrets above them, young men idling on motorbikes below. Past Gimja-bet Mariam. Down Arada, people and shops encroaching so far into the road there is little room for vehicles. Out into the open space above the Saturday market. Stepping graceless, watched and watching, out of the car and into a mass of people and noise. 'Enat Alem! Enat Alem!' Mother of the world, mother of the world. But it isn't Mary they're calling on this time, it's Nannyé, enat alem, enat alem, my mother, my all.

Climbing steps and through the narrow gate into her compound is like walking into memory so old it is indistinguishable from dream. But it has different furniture: two das, one of eucalyptus poles and basketwork just to the right of the entrance, and a larger one, of tenting canvas and metal poles, along the back wall. In the latter stand mourners, mostly women, and a man. Enat alem, enat ale-e-e-e-em, the last syllable bouncing in his throat like a pulse. Then, next to the basketwork das – for it is too small to hold them – a half-circle of deacons, white-turbanned, white-shawled, begin to chant. Sistra clash, and then the drum. Dum, da. Dum, da. Enat alem, enat ale-e-e-e-em. My mead, my rich meat. Dum, da. Dum, da. Dum, da.

The coffin was taken ahead into the house and now sits draped in cloth of red and gold at the centre of her living room. We form a dark wall down one side, facing priests and deacons in embellished robes and golden skullcaps on the other. On the floor, beneath the head of the coffin on its stand, sits a flat basket containing offerings of wheat, of holy water, of frankincense. A censer swings, and smoke rises into the corners, around photographs of strangers (for the house is rented now, has been for years), through their careful bouquets of plastic flowers and their best crockery, and the words that followed her all her life, had been among the first she learned to read, unspool into this space she loved so much. Thou wast named Beloved Woman, oh blessed among women. Thou art the second chamber – The churchmen, young, sweet-voiced, a bit shambolic, clearly bored, conduct whispered conferences between each section of the service, in which gospel follows covenant follows the Book of the Praise of Mary in straggling close harmony. Few apart from the clergy speak Ge'ez, but even I, who am moved by this meeting of high church ritual and domestic intimacy, recognise some of it. Weletè Amanuel, daughter of Amanuel, return now to the country from whence you came. Our Father, who art in heaven.

And then they are lifting her up, and out. No family are allowed to touch the bier. This is an honour accorded only to priests and to the wives of priests who have remained true to their vows, but it makes her children and grandchildren, in an instant, peripheral. They don't go far – they set her down in the yard, and gather round again. This time there is no space next to her for anyone but clergy. Their umbrellas, white and gold robes, processional cross, blink in the sunshine. Her trees tower above us, eucalyptus leaves and the pale beginnings of eucalyptus fruits are scattered across the ground. A purple morning glory has twisted itself about the electrical cables between gate and house, but of her garden nothing else remains. The keening reasserts itself, and the drum.

Out of her gate, through a hubbub of bodies, and back onto the pickup. But almost as soon as everyone sets off, following the boat of the truck and its umbrella sails into the Saturday market, a horn sounds, five times, an old old sound, and all halt. A few of the umbrellas are unusual in shape, conical on top, with Ba'ata picked out in link-metal on the straight-falling velvet sides. These dibab only ever come out with the Ark of the Covenant on holidays, or when clergy die. Another honour. The deacons range themselves in facing lines, and, against a backdrop of legless, faceless shop mannequins, of garish bolts of cloth and curious children, begin to dance. The drumbeat is regular now. A step, a shallow bend of the knees, and a small move sideways per beat, so the lines of white turbans dip and rise, dip and rise. The dance of David, of reeds in water.

The horn again. The procession moves on, streaming through the market, halting work, getting in the way of garris and buses, thus announcing her passing. There's lots of chat. Catching up on gossip, marriages, divorces, deaths, who finally found a job, who has escaped to America. And stop. Some mourners hold up framed photographs of Nannyé in youth, others have draped dresses we brought from Addis, her favourite dresses, over their backs.

At each stop the churchmen make their enclave, the chief mourners a parallel one. They declaim in scratched voices poems of who she was. Wife, mother – imposed roles, unquestioned and in her time unquestionable; passive, in a way, however fully inhabited and lovingly dispatched. She gave her daughters and granddaughters the chance of something different, and in making that gift separated them from her in fundamental ways. The mourners speak of her kindness, her generosity, her many griefs. (She herself could not begin to cope with funerals like this, so when each adult child died she was told of their burial after the fact.) Of how she never got to Tana Qirqos, but in her late eighties made a pilgrimage to Lalibela in the north, where in the thirteenth century vaulting rock churches were hewn out of the mountainsides to make a new Jerusalem, and even to Jerusalem itself, walking to Golgotha on swollen feet and raging because she was not strong enough to climb down to Mary's tomb.

In her last ill months she began to beg to be taken, please, Lord, is it not time? whilst also being greatly afraid. She worried she had not fasted enough, had not prayed enough, would be found wanting. She asked not to be taken in the rainy season, so she did not have to be interred in mud and storms. That prayer has been abundantly granted. The sky is a radiant cloud-scudded blue, and the hills and fields bake in the sun.

We arrive at a sharp bend in the cobbles. The land has opened out. In front of us, beyond dusty figs, rubbish-filled ditches and struggling olive trees, it plunges down to the Qeha, then rises again into mountains and Gonderoch Mariam. These are her streets, her landscape, her home. Even though she has been away for so many years, the white-veiled stream of people flows as far as I can see.

The pickup chugs and tips up a small rise, stops under an unfinished arch. Old old trees, the glory of churches, sough over walls painted so they shine like plastic. I've been told that when, a few years ago, they knocked down my grandfather's church in order to build this one the sanctuary was so well made, the stones so tightly fitted together, it took them ages. Good, I thought, indignant on my grandmother's behalf, thinking of all the work and pain that did for mortar. Good. I'm glad it gave them trouble. This structure, recently consecrated a cathedral, is bombastic, ugly. Bigger was better, as big as possible, overshadowing all others – then again, when was that ever not the case, where Ba'ata was concerned?

A long blast on the horn. Another, and another. The truck rocks in, up, past the main church, and Nannyé is unloaded onto a patch of bare earth between the bell tower and the humble building in which she was married. Again the clergy in their vivid plumage close in around her. This is the last of the seven stops; here they recite the seventh of the seven chapters of the Book of the Praise of Mary. They swing a censer into air alive with birdsong. The mourners, plain in comparison, like female birds, make their own circle around an empty space and sing their secular plainchant. Enanney enanney enanney. Enat alem enat alem enat alem. I drag my black scarf up over my head.

Then the churchmen carry her up the steep new steps and into the sanctuary so she can say goodbye to it. In the yard there is a distinct slackening. A child cries on its mother's back. Someone answers a mobile phone. There are vultures in the trees, dark weights on branches that look too frail to hold them.

When they come out again they are ringing a bell, a continuous ringing that jangles already jangled nerves. They have lit long wax tapers, so flames accompany her circuit, once around the balcony, maregn Christos, maregn Christos, Christ have mercy, Christ have mercy, Christ have mercy, forty-one times, and down, into a crescendo of bells, of keening, of plain, heart-scouring weeping. And the drum. It's the end, cry the mourners. It's the end. She is escaping us.

After all this preparation the final step goes unceremoniously fast. She is already interred, and the men are huddled around the family tomb, holding everyone else at a distance.

Finally, a last long blast of the horn. Come, say your goodbyes. The entrance is narrow, those going in have to flatten themselves against the wall to make space for those going out. She has been placed in the space next to her husband, but they haven't quite been able to close the door, so a tall white wreath is propped up against it, keeping things decent. There is pushing and shoving and wailing; all I can think is, no quiet, no quiet, give us quiet, give me quiet, give her quiet.

A couple of days later, early in the morning, I returned alone. I watched as a man climbed to the top of the bell tower and settled down to read. Above him a kite ascended, trusting itself to the air. Crows cawed demands, and the doves answered. Figures in white sat against tree trunks mouthing verses from fat little books. Women walked bowing up to a workmanlike priest who sat under the balcony handing out holy water and blessing them with his cross, briskly tapping head, back, belly, anywhere the concern was greatest. They kissed the threshold, removed their shoes, covered their hair, and entered the church, as my grandmother had done, morning after morning, for decades. I went down into the tomb.

Torn spiderwebs hung from the roof and a weak light crept through the windows. She had been moved and now lay in a larger space, next to Molla and just above Teklé and Alemitu. Here she fitted, exactly. From inside the church a voice rose, and a drum began to beat, quiet and slow. A cockerel crowed. And I took my leave.

CHRONOLOGY

HISTORY	YETEMEGNU (NANNYÉ)
1775 – Establishment of Ba'ata	
1866 – Tewodros II sacks Gondar	
1886 – Addis Ababa established	
1888 – Gondar burned by the Mahdi army	
3 November 1889 – Menelik II crowned emperor, and Taitu Bitul his empress	
23 July 1892 – Tafari Mekonnen (later Emperor Hailè Selassie I) born	
	1893 (1886 in Ethiopian calendar*) – Tsega Teshalè born
1896 – Battle of Adwa	
	1911/12 (1904 in Ethiopian calendar) – Tsega, aged eighteen, arrives in Gondar with Memhir Hiruy
12 December 1913 – Death of Menelik II	

* The Ethiopian calendar has thirteen months (the thirteenth, Pagumé, has five or six days depending on if it is a leap year). New Year's Day equates to 11 or 12 September in the Gregorian calendar; the seven- or eight-year difference arises because of differing original calculations for year zero, the date of the Annunciation. *The Wife's Tale* proceeds according to the Ethiopian year, months and seasons, except for the dates at the beginning of each book, which are Gregorian. I am indebted to https://www.funaba.org/cc for conversions between the Ethiopian and Gregorian calendars.

1916 – Lij Iyasu, Menelik's grandson, deposed, and Zewditu, Menelik II's daughter, crowned empress; Tafari Mekonnen becomes ras

6 January (Tahsas 28, the feast of Ammanuel) ?1916* – Yetemegnu Mekonnen† (Nannyé) born in her maternal grandmother's house, close to Ba'ata church

28 September 1923 – Ethiopia admitted to League of Nations; slave trade banned

? 1924 – Yetemegnu marries Tsega in Ba'ata church
2 November 1926 (Tiqimt 23, 1919) – Tsega awarded aleqa-ship of Ba'ata church

October 1928 – Ras Tafari crowned Negus (King) Tafari of Gondar
2 April 1930 – Empress Zewditu dies
2 November 1930 – Coronation of Negus Tafari Mekonnen, who becomes Emperor Hailè Selassie I (his baptismal name)

November 1930 – Tsega present at Hailè Selassie's coronation. Alemitu born
? – Baby girl born to Yetemegnu. She dies, unnamed, just before her christening day
July ?1935 – Edemariam born

3 October 1935 – Italy invades Ethiopia
1 April 1936 – Gondar falls
2 May 1936 – Hailè Selassie leaves Ethiopia
5 May 1936 – Italian forces enter Addis Ababa

---

* Yetemegnu was not unusual in carefully remembering birth times, days, months, saints' days, but not being specific about years. Formal birth certificates were not used in Ethiopia at the time.

† Ethiopian naming convention is that the surname is the father's first name, i.e. Yetemegnu Mekonnen, because her father was called Mekonnen Yilma (and his father Yilma Woldè-Selassie). A woman does not change her surname upon marriage, though in some places the groom's family will give her a new first name. Most born into the Orthodox Christian faith have both a daily name and a baptismal name, given at their christening.

19 February (Yekatit 12) 1937 –
Massacres in Addis Ababa, ordered
by Viceroy Graziani. All
intellectuals killed. Followed by
execution of 297 monks at Debrè
Libanos on 21 May, and 1,500
people shot as soothsayers

1939 – Abunè Yohannes replaces
Abunè Abraham

20 January 1941 – Emperor Hailè
Selassie, Ras Kassa and British
forces enter Ethiopia

5 May 1941 – Emperor Hailè Selassie
returns to Addis Ababa

25 November 1941 – Battle of
Gondar

2 December 1941 – The British begin
dismantling Italian factories; in less
than a year they take 80 per cent of
industrial assets

1942 – Asratè Kassa, only surviving
son of Ras Kassa, made deputy
governor-general of Wollo,
Begemdir and Semien

1942 – Mitiku Jemberé, later Abunè
Theophilos, made dean of the new
cathedral

May 1945 – Ras Imru made
governor-general of Begemdir

14 January 1951 – Etchegé Gebrè-
Giorgis becomes Abunè Basilios,
the first Ethiopian archbishop.
Mitiku Jemberé becomes his regent
and deputy

1952/53 – Asratè Kassa returns as
governor-general of Begemdir

? Late 1937 – Birth of Yohannes

? 1939/40 – Teklé born

December 1940 – Dedication of
Ba'ata church

1941 – Yetemegnu, pregnant with
Molla and carrying Teklé, escapes
on a mule to her father's house in
Atakilt Giorgis; Tsega goes to war in
support of Ras Birru

? 1941/42 – Molla born in Atakilt
Giorgis

1942 – Tsega leaves the battlefield to
take his family to Denqez

c. September 1942 – The family
return to Gondar; Yohannes dies
two months later

1942 – Tsega promoted to judge by
Hailè Selassie

1948 – Tsega gets Ba'ata back

? March/April 1953 – Tsega
imprisoned

1953 – Yetemegnu miscarries her tenth child. Goes to Addis to petition for Tsega's release

16 December 1953 (Tahsas 7, 1946) – Tsega dies

1955 – Silver jubilee of Hailè Selassie's coronation

1956 – Molla and Teklé enter Fasiladas boarding school in Gondar; Tiruworq and Zenna sent to Princess Tenagnewerq school, also in Gondar, then to Harar

26 October 1957 – Asratè Kassa promoted to vice-president of the Senate

28 June 1959 – Abunè Basilios consecrated as first Ethiopian patriarch. Church achieves autonomy from Alexandria

December 1960 – Attempted coup led by Mengistu and Germamé Neway

1961 – Edemariam leaves for Canada

August 1969 – Molla goes to Russia, via Khartoum

1971 – Edemariam and his wife Frances arrive from Canada

1971 – Abunè Theophilos becomes patriarch of Ethiopian Orthodox Church. Asratè Kassa appointed president of the Crown Council

1972 – Hailè Selassie's eightieth birthday

1973 – Wello famine

January 1974 – Revolution begins with army revolt in Negellé Borena

1973–74 – New house built for Yetemegnu according to plans by Rev. Harold Lester, Frances's father

12 September 1974 – Hailè Selassie deposed

24 November 1974 – Derg announces it has executed its chairman, and sixty other dignitaries

March 1975 – Nationalisation of all rural land; abolition of the monarchy

July 1975 – Nationalisation of all urban land and 'surplus housing'; church also loses its main economic assets

27 August 1975 – Emperor Hailè Selassie dies

May 1976 – Abunè Theophilos arrested

3 February 1977 – Launch of the Red Terror

14 July 1979 – Abunè Theophilos executed

1984 – Worst famine in Ethiopia in over ninety years affects eight million, kills one million; tenth anniversary of the revolution

1989 – Royal women released from prison

May 1991 – Mengistu flees; EPRDF enter Addis Ababa

5 November 2000 – Emperor Hailè Selassie buried

8 December 2006 – Mengistu convicted *in absentia* of genocide and crimes against humanity

1977 – Molla returns from Moscow to work as a doctor

January 2001 – Alemitu dies

October 2001 – Yetemegnu makes pilgrimage to Jerusalem

May 2009 – Teklé dies

14 August 2012 – Molla dies

21 December (Tahsas 12) 2013 – Yetemegnu dies

GLOSSARY

abet – 'Yes, sir!'; 'At your service'; also, 'Abet! Abet!' means 'Justice! Justice!'

amolé – salt bars, historically used as currency

araqi – arrack, alcoholic liquor

awiy; wiy – an exclamation, roughly 'Oh no!'

ayzosh/ayzoh/ayzot – take heart ('ayzosh' when addressed to a female, 'ayzoh' when to a male, 'ayzot' when to an elder or someone higher on the social scale)

berberé – dried, ground and spiced chillis

Betè Israel – literally, house of Israel – Ethiopian Jews

bethlehem – a small building in the church grounds where nuns prepare eucharistic bread

birilé – small glass drinking flask with a round base and narrow neck, used for drinking mead

birr – literally silver, but also Ethiopia's official currency

chat – leaves chewed for their mildly narcotic effect

das – a temporary shelter of wooden poles, covered in leaves and branches. Used for events such as weddings and funerals

dibab – a type of ceremonial umbrella

dirsanè – homily

enaté – my mother

erè – exclamation

THE WIFE'S TALE

finjal – small, handle-less porcelain coffee cup

gabi – a particularly heavy shemma (q.v.), effectively a blanket

gan – large pottery jar, like an amphora

garri – two-wheeled, horse-drawn cart

gasha – shield, bulwark; also about forty hectares of land

ghibbi – compound or courtyard; also castle complex

ihité – my sister

ilbet – a smooth, spiced sauce of broad-bean flour

ililta – ululation/joy cries

injera – flat sourdough bread made with teff flour

jendi – a blanket of hide

kahinat – clergy

kebero – a drum

kosso – a tree, parts of which are an anthelmintic (against tapeworm)

lijé – my child

madiga – very large pottery jar, usually for carrying water; also a large hide sack capable of holding up to 50kg

masinqo – a one-stringed spike-fiddle

masqal – cross; also, as Masqal, the feast of the True Cross, 27 September (Meskerem 17 in Ethiopian calendar)

meqdes – in a church, the sanctuary, where the altar is located

mesob – a table of woven straw with a conical base and a round top, usually the size of one injera (q.v)

netela – a shawl of thin cotton muslin

qiné – church poetry, sacred hymn

shemma – a shawl, which can be light single-ply muslin – called a netela (q.v.) – or thicker (double- or triple-ply)

shifta – bandit/rebel/guerrilla

shirro – a flour of dried, spiced chickpeas and split peas which becomes a smooth sauce when cooked; a staple meal, especially for poorer people

siljo – a sauce of spices, fermented barley and sorghum

tabot – the Ark of the Covenant, a copy of which is found at the centre/at the altar of every church

teff – the seeds of Williams' love grass/annual bunch grass. Used to make injera (q.v.)

tej – mead

teklil wedding – a church wedding, and indissoluble (other forms of wedding allowed for divorce)

tella – beer

terèt – a folk tale

woizero – respectful title for a woman: Madam, lady, Mrs

yelam berèt – literally a cow's byre, but part of a call and response that transcends literal meaning

zar – a spirit; a possession cult or therapeutic society

Ranks and titles

abba – father; also Father, as in a title for priests and monks

abunè – archbishop, patriarch

aleqa – administrative leader of a large monastery or an important church

bitwoded – beloved [of the realm]

dejazmatch – general, commander of the gate; title just below ras (q.v.)

fitawrari – commander of the vanguard; title just below dejazmatch (q.v.)

Jan Hoy – a title by which to address the emperor

liqè-kahinat – head of the clergy of a province or a large area

meto-aleqa – commander of a hundred

negus – king

negusè-negest – king of kings, or emperor

qengazmatch – commander of the right flank

ras – head of an army; highest title below king

ACKNOWLEDGEMENTS

While I may have grown up in Ethiopia, there were many reasons (not least the revolution and its attempts to erase and recast the past) why things familiar to my grandmother were unknown to me, so I needed years of reading before I could begin to understand the world in which she grew up. I can only mention a fraction of the books and articles here, but Bahru Zewdé's *A History of Modern Ethiopia 1855–1974* was invaluable, as was his work on the Gondar census of 1930–31 and his commentaries on the revolution, in particular his insights about what it did to language, and the dark jokes people told. Solomon Getahun's *History of the City of Gondar* was foundational, as was Adna Abejè's BA thesis on 'Ba'ata Church in Gondar (1775–1968)', written at Addis Ababa University in 1990. *Gondaré Begashaw* by Gerima Taferè provided eyewitness accounts, as did Blatengeta Mahitemè-Selassie Woldè-Mesqel's *Zikrè-neger* and Evelyn Waugh's *Waugh in Abyssinia* and *Remote People*. Despite his opinions and his mockery, Waugh was a good noticer. I spent whole days, weeks, even, in the company of Thomas Lieper Kane's two-volume Amharic–English dictionary. *The Encyclopaedia Aethiopica*, Vols 1–5 (Harrassowitz Verlag) is a remarkable undertaking, as is, in a different way, Simon D. Messing's huge unpublished 1957 PhD 'The Highland-Plateau Amhara of Ethiopia', from which the descriptions of agricultural work assigned to each month are taken. Messing's transcriptions of the familiar stories of Aleqa Gebrè-Hanna were very useful, as were his investigations into zar culture; Ronald A. Reminick and Wolf Leslau were also helpful with this, especially with regard to zar language. Sylvia Pankhurst's weekly newspaper the *Ethiopia Observer* was invaluable, as were book after article after book by her son, the Ethiopianist Richard Pankhurst; as was, in turn, the book *Gender, Development and Identity in Ethiopia* by *his* daughter, Helen Pankhurst. Selamawit Mecca's insightful study of the hagiographies of Ethiopian female saints (with special reference to Gedlè Christos Semra and Gedlè Feqertè Christos) was very useful.

As I do not understand Ge'ez, the church language (which relates to Amharic rather as Latin does to Italian), I was grateful for E.A. Wallis Budge's translations of *The Legends of Our Lady Mary the Perpetual Virgin and Her Mother Hanna*; *One Hundred and Ten Miracles of Our Lady Mary*; *The Book of the Saints of the Ethiopian Church*; *The Book of the Praise of Mary*; *The Virgin's Lyre*; and *The Queen of Sheba and Her Only Son Menyelik* (otherwise known as *The Kibrè Negest*, or *The Book of the Glory of Kings*). The quotations that tell the story of Mary's life, the prayers of Ruphael and of the praise of Mary are all from these books.

Donald Levine's *Wax and Gold* illuminated attitudes and assumptions so familiar I hadn't known I was surrounded by them: the phrase 'the dance of David, of reeds in water' I owe to him. Alberto Sbacchi's *Ethiopia Under Mussolini: Fascism and the Colonial Experience* was very informative, and I am grateful for Reidulf Knut Molvaer's fascinating study *Tradition and Change in Ethiopia: Social and Cultural Life as Reflected in Amharic Fictional Literature*; to his translation of the chronicles of Zewditu, and to his synopsis of *Dubb-Ida* by Balambaras Mahitemè-Selassie Woldè-Mesqel, a detailed account of the characters and events in the abortive coup of 1960. Richard Greenfield's account in *Ethiopia: A New Political History* was vital too, especially his transcriptions of radio broadcasts. BBC journalist Blair Thomson's *Ethiopia: The Country that Cut Off its Head* provided a similarly vivid day-to-day, even hour-to-hour sense of seismic events; the insight that that year's Masqal bonfire collapsed in on itself is his. I am grateful for Philip Marsden's biography of Tewodros, and also for a specific line from his *The Chains of Heaven*, which appears almost unchanged, it so exactly describes what I remember: 'not everyone lying by the side of the road was sleeping'. I am indebted to Marina Warner, whose sympathetic and learned account, in *Alone of All Her Sex*, of how the worship and iconography of Mary is simultaneously an ideal, a comfort and a subjugation, helped me to understand fundamental things about my grandmother and what she shared with women across the Coptic and Catholic worlds; the phrase 'surrounded by more frescoes which, because Yetemegnu could not read, were in effect her Bible' is a quotation and a tribute. Colm Tóibín's *The Testament of Mary*, for leading me to Warner, and for the freedom of the voice. And Michael Ondaatje, whose memoir *Running in the Family* (and his biographical novel *Coming Through Slaughter*) suggested, nearly two decades ago, a possible way in which to approach what is essentially a subjective oral history.

And, as such, my greatest debts are to everyone who told me stories. My sincere thanks to Felegush Mekonnen and her daughters Serawit and Seged Getahun, to Seyoum Getahun, Adanè Jemberu and Neberu Ayalew, Qes Addisé Mekonnen, Migib Tadessè and Kassa Worku, Dr Fisseha Gebrè-Selassie and Abebech Yosef, Derejè Seteng and Tigist Hailu, Qes Dawit Beqalu, Aderajew Asfaw, Habtamu Teklé, Qes Girma Mengistu, Robel Yohannes, Abebè Gedefaw, Dagm, Tewodros and Mesfin, all in Gondar. Professor Bahru Zewdé, Mamo Hailé, Damtew Bizuneh and

Abebech in Addis Ababa, and Abba Teklè-mariam in Debrè Libanos. HIH Prince Be'edè-mariam Mekonnen, Princes Asfè-Wossen Asratè Kassa and Mulugeta Asratè Kassa – all shared memories and/or guidance and material support.

Thanks to Tim Rostron, John Pearce, Christopher Clapham, Lisa Dwan, Karolina Sutton, David Levene, Martin Orwin, John Binns, Kumlachew Muluneh, Laura Thomas, Kieron Humphrey, Andy Beckett, Paula Cocozza, Ariane Koek, Benjamin Markovits, Joanna Kavenna, Leah McLaren, Kazvare Knox and Sarah Habershon for everything from the first germ of encouragement to design and orthographical advice to a room to work in. To the librarians at the British Library (especially in the Map Room), at SOAS, at the Bodleian and at the Oriental Institute library, University of Oxford; to Anne Catterall at the Sherardian Library of Plant Taxonomy, University of Oxford, and to Weinishet Behailu and Genet Getaneh at the Ethiopian Institute and Museum, Addis Ababa. To *Guardian* colleagues (and my bosses over the years) Ian Katz, Katharine Viner, Becky Gardiner, Charlie English, Sally Weale, Paul Johnson, Clare Margetson and Kira Cochrane for their patience, not least with various leaves of absence. To Robert Young and St Catherine's College, Oxford, who trusted that at some point I might fulfil the promises that won me a scholarship that changed my life. To Steven Pollard, Seamus Perry, Jonathan Wordsworth, Bernard O'Donoghue. To the Royal Society of Literature, and especially Jerwood judges Lucy Hughes-Hallett, Andrew O'Hagan and Jane Ridley, for a prize that provided both a much-needed sum of money and, almost more important, a public vote of confidence in my project. To Ellah Wakatama Allfrey for unstinting encouragement and expert guidance, and to Mary Target. To Susanna Rustin and Annalena McAfee, who took a punt on an unknown non-writer and simply assumed the highest ambitions were possible – a rare and generous gift. To Pat Kavanagh, to whom Annalena introduced me, and who offered a similar trust.

To Nicholas Pearson at Fourth Estate, for whose enthusiasm, generous patience and sensitive editing – and for buying the book in the first place – I will always be grateful. Also to the formidable Robert Lacey and the rest of the Fourth Estate team. Wise Terry Karten, Laura Brown, and everyone at Harper US. Anne Collins, Pamela Murray, and the team at Knopf Canada.

My agents Anne McDermid, Melanie Jackson and Peter Straus, a rock in person as well as in name.

Frances Murray, Ania, Maria, Nicola Annett, Caroline Abraham, Jessica Townsend. But mostly Fiona Cameron (of whose willingness to take on months of 5.30 a.m. starts, and never be a minute late, I am still in awe), Lisa Howard and Teddy.

Michael Hughes, Stephen Sandford, Leo Carey, Louisa Bolch, acute readers and steadfast friends.

Nothing would have been possible without the overwhelming help, love and trust of the Ethiopian side of my family, who, except for Alemitu (who had already passed away), endured years of random and often intrusive questions, and many of whom read the final result: Molla, Teklé, Tiruworq, Zenna, Maré and Tigist Tsega, Alemante Gebrè-Selassie, Wodajie Abebè and Abraham Wubé. Patrick, Endrias, Simon and Hiwoté Molla. Elsabet Mitiku. My heartfelt thanks to my siblings Naomi, who combed every line, Yohannes (whose piece on Buhé in *Harper's Magazine* was a guiding light) and Yodit Edemariam, and to my mother, Frances Lester, who put together the glossary, and whose love of reading began it all. My father, Edemariam Tsega, sat for days and weeks with Nannyé and me, translating questions when my Amharic ran out of steam, and answers when my comprehension did. He helped me organise my trips (including a horse trek to Gonderoch Mariam), climbed with me to the holy springs at Debrè Libanos, even though he was in his late seventies, and led me around Addis, showing me where their various small homes had been. We both wrote books at the same time – he about his father and his father's poetry (*The Life History of Liqè Kahinat Aleqa Tsega Teshalè and His Qineis/Yeliqè Kahnat Aleqa Tsega Teshalè Yehiywot Tarikena Qinewochachew*, published in a trilingual edition by Tsehai Publishers in the US), I about Nannyé; we shared facts and insights, in particular about his father's early life and the poems of his death. His dignity, ambition and refusal ever to give up were a daunting example.

To David Dwan, who always raises my game, and who for a decade, and despite the pressures of three jobs, four cities and a book of his own, has given me unwavering emotional, practical and intellectual support, read the manuscript, helped with titles and pulled me out of the inevitable dips in faith – thank you, so so much.

Thank you to Rahel, our daughter, through whose entire life I have been writing this volume. For inspiration, for insightful interventions, and for the utter joy of her presence.

And thanks above all and for everything to Nannyé, whom I began recording, off and on, twenty years ago, whose voice and point of view I have had the temerity to try to inhabit and whose words often exist in direct translation in my lines, whose laughter and mischief, whose unconditional love we all miss every day. Through the actions detailed in this book, she infinitely expanded our options, and our futures. She also placed an extraordinary and humbling trust in me, and I hope I have at least begun to do justice to her experience.

Any mistakes and misjudgements, however, are mine alone.

ILLUSTRATIONS

ABOUT THE AUTHOR

AIDA EDEMARIAM, who is of dual Ethiopian and Canadian heritage, grew up in Addis Ababa, Ethiopia. She studied English literature at Oxford University and the University of Toronto, and has worked as a journalist in New York, Toronto, and London, where she is currently a senior feature writer and editor for the *Guardian*. She is a recipient of a Royal Society of Literature Jerwood Award for a work of nonfiction in progress and lives in Oxford.